PENGUIN BOOKS

GHOST HUNTERS

Deborah Blum is a professor of science journalism at the University of Wisconsin. She worked as a newspaper science writer for fifteen years, winning the Pulitzer Prize in 1992 for her writing about primate research, which she turned into a book, *The Monkey Wars*. Her other books include *Sex on the Brain* and *Love at Goon Park*. She has written about scientific research for the *Los Angeles Times*, *The New York Times*, *Discover*, *Health*, *Psychology Today*, and *Mother Jones*. She lives in Madison, Wisconsin, with her husband, two sons, and a very large boxer.

DEBORAH BLUM

Ghost Hunters

WILLIAM JAMES AND THE SEARCH FOR
SCIENTIFIC PROOF OF LIFE AFTER DEATH

PENGUIN BOOKS

PENGUIN BOOKS
Published by the Penguin Group
Penguin Group (USA) Inc., 375 Hudson Street, New York, New York 10014, U.S.A.
Penguin Group (Canada), 90 Eglinton Avenue East, Suite 700, Toronto,
Ontario, Canada M4P 2Y3 (a division of Pearson Penguin Canada Inc.)
Penguin Books Ltd, 80 Strand, London WC2R 0RL, England
Penguin Ireland, 25 St Stephen's Green, Dublin 2, Ireland (a division of Penguin Books Ltd)
Penguin Group (Australia), 250 Camberwell Road, Camberwell,
Victoria 3124, Australia (a division of Pearson Australia Group Pty Ltd)
Penguin Books India Pvt Ltd, 11 Community Centre, Panchsheel Park, New Delhi – 110 017, India
Penguin Group (NZ), 67 Apollo Drive, Rosedale, North Shore 0745, Auckland,
New Zealand (a division of Pearson New Zealand Ltd)
Penguin Books (South Africa) (Pty) Ltd, 24 Sturdee Avenue, Rosebank, Johannesburg 2196, South Africa

Penguin Books Ltd, Registered Offices: 80 Strand, London WC2R 0RL, England

First published in the United States of America by The Penguin Press,
a member of Penguin Group (USA) Inc. 2006
Published in Penguin Books 2007

Grateful acknowledgment is made for permission to reprint selections from the following works:
Materials in the archives of the American Society for Psychical Research, Richard Hodgson
Collection and James H. Hyslop Collection. Used by permissions of the American Society for
Psychical Research, Inc. (ASPR), New York, New York.
Letters of William James in the collection of the Houghton Library (bMS Am 1505 and
bMS Am 1938). By permission of the Houghton Library, Harvard University.

THE LIBRARY OF CONGRESS HAS CATALOGED THE HARDCOVER EDITION AS FOLLOWS:
Blum, Deborah
Ghost hunters : William James and the search for scientific proof of life after death / Deborah Blum.
p.   cm.
Includes bibliographical references and index.
ISBN 1-59420-090-4 (hc.)
ISBN 978-0-14-303895-5 (pbk.)
1. James, William, 1842–1910.   2. Parapsychology—History—19th century.
3. Spiritualism—History—19th century.   4. Ghosts.   I. Title.
BF1028.B57   2006
133.909'034—dc22   2006044948

DESIGNED BY AMANDA DEWEY

146119709

To the sisterhood—Darcy, Dawn, and Dana

CONTENTS

Ghost Hunters

PRELUDE

⁓

No one saw the girl die. It was just a little too early, a morning still too dark, first light barely warming the edge of the sky. The night frost yet shimmered on the ground, a faint ghostly silver. It was barely 6:00 a.m. on a late October morning when sixteen-year-old Bertha Huse stepped out into the quiet. Her parents and her sister were still asleep in the small house they shared in Enfield, New Hampshire.

Later, a few of the town folk, those like the blacksmith's wife who were up doing morning chores, recalled seeing the girl go by, tying her bonnet as she went. She was a familiar sight, catching a quick walk before she started her mill job. The blacksmith's wife watched Bertha turn toward the lake. She left a trail of footprints on the frosty dirt road, but they vanished as the sun climbed higher and warmer.

When their child didn't come home, her parents were at first puzzled. Then a little worried. And slowly, as the sky lightened and the morning quickened, they became frantic. They were out calling for her, hunting down the empty road to the shores of the silent lake. As it became obvious that something was really wrong, their friends and neighbors joined in. By day's end, the searchers numbered more than a hundred and fifty people, calling ever louder, rattling through empty thickets and fields, more and more of them glancing toward the polished blank of the water.

The dirt road led to an old Shaker bridge, a structure of wooden boards, without railings, offering a simple passage across Mascoma Lake. The lake was longer than wide, clear gold at its edges, darker at its center where the water deepened, shining like an agate set into the burnished hills. Around the bridge though, the water turned opaque, clotted with weeds growing around the wood, dropping to a black depth of eighteen feet. The searchers found no trace of the girl, not in the woods, in the clear shallows, or around the bridge where the water gleamed as a dark mirror, reflecting only the worried faces.

Finally, the mill owner himself sent for a diver from Boston to explore around the Shaker bridge. The diver, suited against the cold of the lake, slid down, disappearing himself like a stone beneath the surface. Again and again he tried, until his leather suit was sodden and his glass bubble helmet filmed with lake water. But he didn't find the girl either. It was as if Bertha had vanished into the dawn itself. There was nothing—except the nightmare that caused a woman in a nearby town to start screaming in her sleep.

GEORGE AND NELLIE TITUS lived almost five miles away, down lake, in the mill town of Lebanon. Like Bertha, George worked in one of the clothing mills. The workers spread the tale of the lost girl. Everyone knew—and worried—about a missing child. The Tituses talked it over, wondered if they could help.

After supper, Nellie Titus went upstairs and curled into a rocking chair for a nap, pulling a blanket over herself. She was asleep when her husband came up the stairs, but she was sitting almost bolt upright, her hands wringing in her lap and her breath coming in gasps. She screamed. Alarmed, he reached over and shook her awake. Her eyes stayed unfocused. But her voice came sharp and unhappy. "Oh, George," she said. "Why did you wake me?" A little deeper into the dream, she insisted, she could have told him where that girl was. He shook his head. It was a nightmare, he said, you were screaming. "Don't wake me again," she answered, almost pleading. "No matter what I do in my sleep."

The dreams haunted them both for another day. Bertha Huse had disappeared on a Monday morning. That Thursday, George Titus gave up on sleep at 1:00 a.m. He lit a lamp and turned toward his wife. She was shivering now, teeth chattering. "Are you cold, Nellie?" he asked as he pulled the quilt up. "Oh, oh, I am awfully cold," she answered, and then she slid silently down into the bed. When she woke, the dawn was once again breaking against the blackness of the night. We have to go to the Shaker bridge, she told her husband.

He was almost too exhausted to argue. And anyway, she was scaring him. Titus went to the stable where he kept a horse and told the stable master, almost casually, that his wife had dreamed the answer to the missing girl. The man laughed, but George Titus shook his head and recounted what she had told him. That Bertha Huse had first appeared on the bridge, hesitating about which way to go. She balanced on a frost-slick board, looking back toward the village. Her foot slipped, and she went backward into the lake. It was about at this point in the dream, he thought, that his wife had begun shaking with cold.

The Tituses drove their buggy onto the bridge. They were more than halfway before she asked to stop. Nellie Titus walked to the edge. "George," she said, pointing down the bridge supports. "She is right *there.*" He could see nothing but the polished surface of the water, black and glossy as obsidian. He stared at his watery reflection. There was one other thing that he knew, which he hadn't told his friend at the stable. Nellie was the granddaughter of a known psychic. She didn't claim to be one herself. But she'd had these haunted dreams before. They always left her physically ill. She hated them. They came anyway, slipping past her defenses as she slept.

The couple tracked down George Whitney, the mill manager, who was heading up the search team, and he—frustrated and desperate—called in the diver. The diver was a slim, wiry Irishman named Brian Sullivan. He thought they were all crazy. Sullivan worked for the Boston Tow-Boat Company, and he'd been diving for bodies for years. Bodies were sometimes hard to find, that's all. That was reality. Dreams weren't. Whitney told him that he didn't believe in such dreams either. They were both rational men, he assured Sullivan; they both agreed that ghosts did not really creep into dreams.

But Whitney knew people in the village who were less "rational." There were those who did believe in visiting spirits. And he told Sullivan, "As long as we have started to do all we could to recover the body, we ought to at least give this woman a chance."

Sullivan continued arguing. He wanted to send to Boston for blasting powder and set off an explosion that would shake the girl's body loose from the lake bottom, if it was there. He told Whitney that he was willing to take orders—but he did not want make a fool of himself, splashing in and out of the lake while Mrs. Titus spouted mystical nonsense and pointed at assorted dream spots. Whitney conceded the point. He had no intention of looking ridiculous either. They would give her one chance only.

Nellie Titus walked on the bridge again and stood where the horse had first stopped. She shook her head. No, she said, not quite right. She walked a few more feet and leaned forward a little, staring down into the lake. "Here," she said. And she told the diver that the girl was wedged upside down in the wooden structure of the bridge, but that one foot was sticking up, still in its rubber overshoe.

Sullivan put on his dive suit and dropped a guideline, tied to a sinker, down into the water. The lake was so dark that once underwater, he could see nothing. He slid down ten feet into the black of the water, feeling his way down the bridge structure. Something bumped his face. He fumbled to feel it. It was a foot. His hand slid over the rubber boot and down a leg, and he began to pull. He tied the guideline round the body and scrambled up to give the news.

"She's here," he shouted as his head, monsterlike in its glass bubble, broke the surface.

Whitney shouted back, "I know."

Even before Sullivan freed the body, the girl's bonnet had tugged loose, floating upward until it hung just below the surface, like a sodden flower.

Later, at the inn tavern in Lebanon, Sullivan told his story to a fascinated crowd. Wasn't he afraid, someone asked, when he was down there in the pitch-black with the girl's body bumping his face? No, he answered. "It is my business to recover bodies in the water, and I am not afraid of them." But he feared something else—or rather, someone else. He was unnerved by how precisely Nellie Titus had described the location of the body. They wouldn't have found it without her, he said; he couldn't even see the body when he was floating next to it. What power did she have that a dead girl walked in her dreams? "In this instance, I was afraid of the woman on the bridge."

The woman on the bridge drew another audience, one that saw much more than a fearful oddity. Her story was investigated and then published by a surprisingly elite group of scientists and philosophers. They viewed it as evidence, a data point in an ambitious—and professionally risky—scientific endeavor to prove the existence of life after death.

One philosopher, a supporter of the endeavor, recalled being given a very direct warning against the effort, having a colleague direct his attention to a "once eminent chemist" with the comment that the man "*had* been a brilliant scientist but that recently he had unfortunately gone off his head" and begun investigating the supernatural. And yet investigating the supernatural—using the techniques of science to explore the occult—was exactly what William James and his friends wanted to do, in fact, considered a necessary part of science and its mission to understand the world. It was James who wrote up the story of the drowned girl for publication, neatly sifting the evidence for answers. His report cited interviews with everyone from the Boston diver to George Titus, from the mill owner to the blacksmith's wife. In his conclusion, James ruled out fraud, any chance that Mrs. Titus had been lurking around the lake in the frosty dawn. He ruled out any relationship at all between the dead girl and the woman who helped find her. His report included the diver's first reaction—"I was *stunned*"—and the Huse family's baffled gratitude for help from a stranger with a "God-given power."

Like the once-acclaimed chemist, William James had a reputation to lose. He was a heralded professor at Harvard University, the author of the most respected text on psychology yet published, a founder of the American Psychological Association. He'd developed an innovative approach to philosophy, rattling elitist traditions by linking everyday, real-life experience to intellectual exploration. He'd become famous even beyond academic circles, almost as well known as his novelist brother, Henry James Jr. The timing, in terms of personal credibility, was absolutely terrible. He said what he thought anyway: "My own view of the Titus case consequently is that it is a decidedly solid document in favor of the admission of a supernormal faculty of seership."

James and his companions in this scientific ghost hunt were famed for their intellectual brilliance—their intellectual courage gained them less admiration. Yet they possessed both qualities in abundance. James's fellow ghost hunters included the codiscoverer of the theory of evolution, a physiologist from France who would win the Nobel Prize in Medicine, an Australian who became a founding member of the American Anthropological Society, a female mathematician who became principal of Cambridge University's first college for women, a pioneer in British utilitarian philosophy, and a trio of respected physicists.

All of them had reputations that would suffer as a consequence, and all of them, like William James, would refuse to abandon the search. For all the risks, the work had its rewards as well—participating in what would become the best ghost hunt in the history of science, conducting studies that would frequently startle and sometimes convert skeptics into believers. All of them believed that they were working toward an understanding of life that could help bridge the chasms between science and faith, between what people see and what they imagine.

"Hardly, as yet, has the surface of the facts called 'psychic' begun to be scratched for scientific purposes. It is through following these facts, I am persuaded, that the greatest scientific conquests of the coming generation will be achieved," James would write. The requirements for achieving that goal were both simple and almost impossibly complex. They included patience, a belief in infinite possibilities—and a willingness to accept that a dead girl could rise, as ephemeral and as real as mist, above the black lake of a dream.

1

⁓

THE NIGHT SIDE

$\mathcal{W}$ ILLIAM JAMES sensed his own life growing short that year of Nellie Titus's haunted dreams, having celebrated his fifty-sixth birthday the previous January. His cropped dark hair and short beard were rimmed with gray. Smile lines radiated like sunbursts around his eyes. But his eyes remained sharply blue, his body wiry and compact, and his movements still hummed with the intensity of his youth.

James had never been a physically imposing man. He stood just five feet eight inches tall, fine-boned, and high-strung. As a young medical student, with a barely suppressed preference for art, he'd sketched himself as a dreamer—sharp cheekbones, sensitive mouth, flowing hair, drooping mustache, shadowed eyes. Over the years, he'd shed the soft look but not the inner, idiosyncratic dreamer. He found upright Victorian tailoring uncomfortable, preferring to dress in tweed jackets with checked trousers, soft shirts, and thick, cushioned shoes. To the end, he found predictability dull, once complaining that spending time with one of his more methodical Harvard colleagues was "like being in the dentist's chair."

Arguing with a fellow researcher about the importance of the paranormal, James coolly reminded his colleague that science—the nineteenth-century powerhouse behind the electric light and the dynamo, the tele-graph and the telephone—could be both arrogant and wrong. James wrote to *Science* magazine, a journal devoted to upholding the research ethic, that he loathed the reverential use of the word "scientist . . . it suggests to me the priggish, sectarian view of science, as something against religion, against sentiment," even against real-life experience.

His approach kept him often at odds with the views of his academic col-leagues, whom he sometimes referred to as the "orthodoxers." But unortho-doxy came to him naturally. He'd grown up in an almost dazzlingly unstable household, and he recognized, intuitively, the cultural instability of his times. The era was one of intense moral imbalance—religion apparently under siege from science, technology seemingly rewriting the laws of reality. Finding some balance, a way to make sense of existence in the changing world, seemed to him an imperative.

So James refused to accept the easy explanations and absolutes dictated by either side. He went further than that, even, deliberately challenging the traditions of both science and religion by choosing to explore the super-natural, to investigate and even support tales like that of the woman on the bridge.

His work on all fronts—psychology, philosophy, psychic—confirmed his own inclination, that the most important lessons might be learned in the most unexpected places.

FROM CHILDHOOD, William James breathed the air of pure, undiluted unpredictability. It was the atmosphere of life with his sometimes brilliant, always erratic father, Henry James Sr. Born into a wealthy New York fam-ily, married to a wealthy woman, the elder James had enough income to indulge his intellectual quests—and the will to do so. Henry's determina-tion had been honed by his own moralistic father and disciplined mother. But he'd been shaped even more by a horrific accident when he was teenager—and by his sense of the dark supernatural forces behind it.

At the time of the accident, in 1824, Henry was thirteen years old, a student at the privileged Albany Academy in New York. In science classes, the boys were studying principles of flight. In one series of experiments, they ignited paper balloons, dipped in turpentine, and watched the fire's heat drive the balls upward. When burning remnants fell to the ground, the boys stamped them out. One day, however, an errant wind blew a balloon into a nearby stable. Sparks sizzled in the hay. Ever impetuous, young Henry James dashed into the building to beat out the fire. Instead, the rising flames set his trousers ablaze, charring his right leg almost to the bone. Surgeons amputated it below the knee, a torturous operation that involved sawing through skin and muscle and then snapping the bone in two. Even so, the damaged leg continued to fester. A year later, the doctors sawed off another section of leg, now spotted black with gangrene. This time the cut was above the knee.

As a result of the accident and surgeries, Henry was bedridden for more than three years. He became increasingly dependent on the alcohol he drank to ease the pain. His father worried over him. But his mother, Catherine, refused to allow pity, or self-pity, into the sickroom. If Henry's father showed too much sympathy, she kept him away from his son. She was a woman gentle in manner but powerful in faith. It was God's will, according to her beliefs, and she wanted her son to find strength in the Lord's sure judgments.

Trapped and miserable, the boy turned to anger instead. And fear. In the years to come, Henry would divide his life into the magical, sunlit childhood before God had ruined his leg with "his dread hand" and the bitter years that followed. The once happy child would grow estranged from his parents. He became a drinker, a gambler, a wealthy drifter—and a man in search of protective cover. Out there, he knew, was a deity on the prowl, in search of bloody retribution for sins. And who knew what they might be? Undue sympathy, lack of prudence, a thoughtless scramble into a burning stable? "I am sure," he wrote once, "no childish sinews were ever more strained than mine were in wrestling with the subtle terror of His name."

Henry James Sr. wrestled also with the demons of depression, a tendency that his children would inherit. Even after he was an established man, with a wife and family, he would describe his life as "not by any means a victory, but simply a battle." He slowly overcame his alcoholism, and in his mid-twenties entered the Princeton Theological Seminary. He wanted to know better the mind of that vengeful God: "When the fire burns my incautious finger, I do not blame the fire, and why? Because I feel that fire acts in strict obedience." With the Lord's dreaded hand hovering over his, James lasted at the seminary only a year. He walked away in 1838. But during that year, he befriended the sisters of one of his classmates, and in 1840 he married the elder, Mary Walsh, after convincing her to abandon her allegiance to the Presbyterian Church.

Foreshadowing their life to come, the newlywed Jameses did not purchase a house of their own. Instead Henry and Mary moved into a hotel, the newly built Astor House. Gleaming with polished granite, with more than three hundred rooms, an interior garden, and brilliant flower beds, the five-story Astor House was among the newest and glossiest structures on South Broadway. Guests entered through a Grecian portico and stayed in rooms boasting innovations such as individual locks in doors and running water in bathrooms. The hotel, near the intersection of Broadway and Ann Street, sat across from Mathew Brady's photo studio and P. T. Barnum's newly opened, flag-festooned American Museum. Carriages clattered past, down a promenade designed to echo the elegant thoroughfares of Paris and London. Gas-jet lights glittered along Broadway at night, from its southern tip at the Battery Gardens to its northern boundary, where it merged into the country roads.

The Jameses lived at Astor House for more than two years. Their oldest son, William, was born in the hotel on January 11, 1842. Within the year, they moved into a house in Washington Square, and their second son, Henry Jr., was born in 1843.

To his diary, James Sr. confessed that he wished God would take Willy and Harry before they were old enough to become sinners. No one stayed untarnished for long, and the Lord's vengeance could only be an angry breath away. As his depression and his fears rose to choke him, he sold the house and fled to England, towing his wife, sons, and sister-in-law with him.

It was there, while he attempted a water cure, a popular remedy for soaking one's ills away, that a fellow sufferer offered comfort by describing his crisis as a rite of passage. She explained that, according to the eighteenth-century Swedish mystic Emmanuel Swedenborg, James's breakdown was merely a "vastation," a turning point that would lead him to new health and harmony with life.

The elder James hesitated to pursue the idea; his doctors had warned him not to overtax his brain. But finally "I resolved that in spite of the doctors, that instead of standing any longer shivering on the brink, I would boldly plunge into the stream"—out of the dark biblical landscape and into the comforting current of a spiritual vision.

To SCIENTISTS, from the very first, the 1840s sang of achievement—the spreading success of the telegraph, the development of surgical anesthesia, the publication of Charles Darwin's *The Voyage of the HMS Beagle*. From that perspective, Emmanuel Swedenborg was an anachronism, dead since 1772, and a failure, since he was in fact a scientist turned seer.

Born in Stockholm in 1688, the son of a Lutheran bishop, Swedenborg had trained as an engineer with a specialty in metals. As a student, he'd traveled to England in his early twenties to learn from such renowned scientists as Isaac Newton and astronomer Edmund Halley. By 1744 he was comfortably set as an administrator of mining ventures for the Swedish government.

That year, though, while traveling on business, Swedenborg experienced a life-altering vision. He was relaxing at the tavern of his London hotel when, he said, a mist formed before his eyes and then separated into a silvery mass of snakes. As the reptiles coiled around the floor, he caught the outline of a man cloaked by shadows in the corner of the room. The following night, when he returned to the tavern, the mist and the snakes again swathed the room. The figure stepped out of the corner, Swedenborg said, and identified himself as the Lord God. The Swedish scientist left London with a new mission—to explain the real meaning of the Scriptures to the world. God would tell him what to write and give him the gift of far sight.

Swedenborg abandoned his scientific career. He became a wraith of a man with a thin, sharply lined face and sunken eyes. He lived on sugared coffee and cakes, which he believed aided a newly fragile digestion. He spent hours in a trancelike state, dream-writing his books on heaven and hell, the spirit world that inhabited our planet and others. For the rest of his life, he spread his version of the gospel, invoking the spirits, which were hidden to all else, by demonstrating his newfound powers.

In his most famous demonstration—investigated by German philosopher Immanuel Kant—Swedenborg disrupted a garden party by suddenly
announcing a firestorm in Stockholm, some three hundred miles away.
Staring at a clear evening sky, Swedenborg described the fire's progress,
street by street and building by building, to the disbelieving company. Two
days later, a messenger from Stockholm confirmed every detail. Kant concluded that the case showed "beyond all possibility of doubt" that Swedenborg did indeed possess an extraordinary visionary gift.

There were other tales, rumors of mysterious insights into lost letters,
revelations about dead relatives. But Swedenborg didn't specialize in showy
clairvoyance. His major teaching, the one that would so catch the imagination of Henry James Sr., was called a "theory of correspondence." It proposed a tangible connection between the material life of this world and the
spirit world, unseen cords that bound inhabitants on both planes together.

Life here, Swedenborg said, is paralleled there, so that our decisions can
influence spirits, their desires influence ours. Everything we touch resonates
in that alternate world. Every action is an interaction with those on the
other side. To be aware of that connection with all God's other realms,
people needed only to let go their earthbound egos. Self-love, self-hate,
self-reflection—all of that creates a kind of blindness, an opaque wall of self
through which we cannot see. And evil spirits can encourage that, Swedenborg said. They can prod us into looking obsessively inward instead of out
into the Lord God's miraculous world. There is no real sin in this universe,
except for perhaps a willful refusal to see the wonders spread before us.

To his followers, science offered nothing to equal Swedenborg's visionary
powers—or his ability to offer comfort in difficult times. Not long after his
death from a stroke, English adherents established the New Church, dedicated to Swedenborgian theology. American followers convened the General
Convention of the New Jerusalem in 1817, also dedicated to his teachings.

In accepting Swedenborg, Henry James Sr. believed that he had found a
new foundation for his life. His fears and sorrows could be blamed on
"ghostly busybodies" seeking to control him. They had infested his life,
these spiritual "vermin revealing themselves in the tumbledown walls of our
old theological hostelry." With that knowledge, he could fight them off and
accept himself as a goodly creation of God. He could follow Swedenborg's
lessons and see failings and personal crises as tests only; accept strengths

and blessings as gifts from the Lord God, without self-congratulation. Therein lay the making of a steady man, a life that much closer to heaven.

Determined to be that man, Henry James Sr. returned with his family to New York. By 1848, he and Mary had two more sons, Wilkie and Bob, and a daughter, Alice. The family moved into a newly built brownstone in Manhattan, just north of Washington Square, along with all the servants that Mary felt necessary to buffer her from domestic duties. The house was also inhabited—as Bob would say resentfully years later—by their father's "idols, the virgin adoration, the faith in miraculous agencies, all . . . the mysteries of the dead."

Swedenborg's unearthly specters encircled the James children's life. When Alice suffered anxiety attacks, her father blamed an intrusive spirit. When William refused to obey his father, the rebellion was undoubtedly made worse by spirit influences. Bob, often the most openly angry of the children, complained that he had never been able to discover a way to achieve simple faith. He had never learned to accept organized religion. The ability to believe "the bed bug priests" demanded comfort and ignorance, and he possessed neither quality.

If there was an afterlife, Bob wrote to his brother William many years later, he did not expect to see their father there. Actually, he hoped not to. There was an ominous possibility that Henry James Sr. would appear trailing Swedenborgian spirits like translucent fish hooked onto a spectral line. Bob James found himself longing for the days of the Crusades, when faith, he thought, must have been as simple as slaughtering a few heathens in the name of God.

BOB'S CRAVING for a simple Christian faith, uncluttered by interfering spirits, defied not just his home life but the times themselves. To the dismay of both scientists and clergymen—united in their distaste for all things superstitious—popular fascination with the spirit world began to spread like a grass fire, driven by that rising sense of moral uncertainty and sparked by events on both sides of the Atlantic. In the year 1848, as the James family settled into New York, a book published in England became an unexpected best seller and—scholars would later say—persuaded generations of readers that believing in ghosts was an acceptable thing:

The footsteps were coming down the hall. They belonged to the son of the house, poor young man. He'd been sickly so long, caught in one of those odd fevers that just kept burning its way back.

He liked to get up, sit with the family in the parlor. It was hard for him, though. His wobbling walks to the sitting room came less and less often. His leather shoes, so little used, began to dry out, despite attempts to polish them soft.

That was the sound. Not that faint groan of a floorboard. The squeak of dry leather. Step, squeak, step, squeak.

The housekeeper was sitting with his sister when they heard him. They hadn't expected him; his parents had sent him from their English country home to Portugal, hoping that the gentler climate would ease the illness. He must have returned just now, in the night. The servant caught up her candle, ran to help. The house was dark, the stairs unlit. Surely he would need light on the way to his room.

Step, creak, step. She couldn't quite catch up with him. There was only the sound of his footsteps moving up the stairs. She hurried after him, heading for his bedroom.

The door was closed. The hall was empty. She opened the door of his room. The candlelight danced across the neatly made bed, the suddenly silent space.

They were almost expecting the telegram when it arrived. The strip of paper, the terse words, telling them that he had died in Lisbon that night, about the time they'd heard those creaking footsteps making their patient way back to his room.

The supernatural account, written out by both servant and mistress, was one of hundreds collected in *The Night Side of Nature,* which turned its author from a little-known writer of children's stories to a celebrity advocate for the occult.

Ghost stories spilled from every page of *The Night Side:* Widowers saw their dead wives walking in the street. Murderers stepped back screaming from the bony brush of their victims' fingers. A father dreamed three times that his son died in a wagon accident, dismissed it as superstition, and changed his mind only on the day that the child fell from a friend's wagon and was crushed by the wheels. A family rented a house so haunted by mysterious voices that their servants left them; their child woke in the night, demanding to know who kept crying.

The author, Catherine Crowe, gathered her tales from friends, from newspaper accounts, other books, letters, and diary excerpts. Crowe had lived in Germany for several years before returning to her native Scotland. She took her book's title from a German term for night at its darkest—on the side of Earth farthest from the sun. Nature's night side in the hours after midnight, when tree branches curve like claws and shadows warp on the wind. But the German term that readers best remembered from her book was *poltergeist.* She was credited for introducing the word into the English language.

The Night Side stayed in print for more than fifty years, convincing thousands of readers that life remained, at its borders, a place of mystery, inexplicable and often terrifying. But Crowe didn't intend her book as merely an anthology of ghost stories. She intended it as a manifesto. "I avow, that in writing this book, I have a higher aim than merely to afford amusement," Crowe insisted in her introduction. "I wish to engage the attention of my readers, because I am satisfied that the opinions I am about to advocate, seriously entertained, would produce very beneficial results."

She wanted to prod scientists into conducting serious investigations of the apparitions she described. She didn't deny that ghost stories were often built on rumor and exaggeration and outright falsehood, but she urged the research community to look beyond, to see the rare flicker of something genuinely supernatural. "If I could only induce a few capable persons, instead of laughing at these things, to look at them, my object would be attained, and I should consider my time well spent."

Crowe flung that hope, that challenge, against a solid wall of scientific hostility. In general, nineteenth-century scientists felt a personal responsibility not to investigate claims of the supernatural but to debunk them out of hand. One British physician had published a book, *An Essay towards a Theory of Apparitions,* which held that ghost sightings floated out of mentally defective brains and that the best cure was "the care of an intelligent physician." Another proposed that ghosts resulted when the brain was excited by an overactive circulatory system and a wild influx of blood into the nervous system. Others simply discussed the human tendency to tumble into hallucinations. None allowed the idea that the ghost seers in question had actually *seen* something.

No wonder, then, that Crowe felt compelled to say that the stories she collected came not from asylums but from respectable citizens relating a moment of shock in otherwise ordinary lives. No wonder her tone was so defensive as she asked for scientific contempt to be replaced with humility. No wonder her demands were so modest: "I assert that whether these manifestations are from heaven or hell, or whether they exist at all, is a question we have a right to ask."

IN THE VERY YEAR that Crowe's book was published, a pair of spirit mediums appeared in the unlikely setting of a New York farm town. In the same sense that *The Night Side of Nature* was considered the most influential publication of its kind, these mediums would come to be hailed as the most revolutionary ghost talkers of their time. In the beginning, though, they were no more than a couple of farm girls, living in Hydesville, New York, which was no more than a scatter of wooden houses, dirt roads, and small farms, about twenty miles from Rochester.

John Fox, his wife, Margaret, and his three daughters had moved to Hydesville the previous year. They had a neat serviceable house, with wood-framed walls and a dirt-floored cellar. They'd heard rumors that it was haunted, nicknamed "the spook house," but the place seemed ordinary enough to them, country quiet and country dark, until the night that Mr. Splitfoot came calling.

The racket began on a spring night in 1848, and it shook the household awake. Something, it seemed, was trying to beat its way out from the timbers. The knocks rattled the rooms, made the floor shiver underfoot. The girls—Leah, sixteen, Margaretta, fourteen, and Kate, eleven—ran screaming into their parents' bedroom.

Night after night, the beat sounded in the walls. Exhausted, tired beyond fear after several weeks, the younger daughter, Kate, stood up in bed and issued a challenge to the rapper in the wood.

"Mr. Splitfoot," she said, giving him a name.
Her voice was high and light, a child's voice.
"Do as I do."
She clapped her hands once.

Came one knock.

Kate clapped twice. Two knocks.

Her sister, Margaretta, snapped her fingers. Again the raps echoed her actions.

Snap. Rap. Snap. Snap. Rap. Rap.

In his book *A History of Spiritualism,* Sir Arthur Conan Doyle, creator of the Sherlock Holmes stories and a devoted believer in the spirit world, tried to re-create the moment: "the earnest, expectant, half-clad occupants with eager, upturned faces," the small pool of candlelight around them, the heavy shadows coiled in the corners of the room, the dead knocking at the walls of the living. "Search all the palaces and chancelleries of 1848, and where will you find a chamber which has made its place in history as secure as this little bedroom of a shack?"

According to the legend of the Fox family, Kate and Maggie next devised a simple code to communicate with the spirits: Two knocks for yes. Silence for no.

"Were you murdered?"

Two knocks.

"Can your murderer be brought to justice?"

No sound.

"Can he be punished by law?"

Silence.

The girls developed a gruesome portrait of the spirit in the house. He was a peddler, killed for his money by previous occupants. His throat had been cut with a kitchen knife, and his body had been dragged, smearing blood, through the buttery, down the stairs, down to where he now lay buried in the earth-floored cellar.

That summer, John Fox and his neighbors started digging in the "spook house" cellar. Five feet down, they found a plank, and beneath the wood a layer of charcoal and quicklime. In the dirt below, they found fragments of bones and hair. They called in a doctor to look over the filthy bits and pieces. He identified them as the pitiful remnants of a human body.

.   .   .

THE STORY OF the Fox sisters was big news in upstate New York and even the surrounding states. It grew to national, even international quality, thanks largely to that master showman Phineas Taylor Barnum.

P. T. Barnum read and enjoyed news accounts of the Fox sisters' amazing powers. He also realized that the girls were wasted in the upstate sticks. They belonged in his American Museum, a marble showcase on lower Broadway, emblazoned with brilliant flags, packed with 600,000 live and dead curiosities—stuffed lions to living fortune-tellers, two-headed men to dancing midgets.

The two younger Fox sisters were the attraction: brown-haired girls from the country—chubby Maggie with her big dark eyes and round face and skinny little Katie with her sharp, birdlike features and restless hands. They were shy, barely educated, simply dressed in their neat dark dresses and white collars. But Barnum was sure visitors would pay to sit down with daughters who talked to the dead. Regular admission to the American Museum was twenty-five cents. To converse with ghosts, people might pay a full dollar. They might even pay two dollars.

"Is the person I inquire about a relative?"
    Two knocks for yes.
    "A near relative?"
    Yes.
    "A man?"
    No answer, meaning no.
    "A woman?"
    Yes.
    "A daughter? A mother? A wife?"
    No answer.
    "A sister?"
    Yes.

The questioner was the novelist James Fenimore Cooper, author of the sprawling American epic *The Last of the Mohicans.* He had come to visit the Fox sisters at Barnum's place, along with other luminaries, including the newspaper man Horace Greeley, editor of the *New York Tribune,* the poet William Cullen Bryant, and Nathaniel Tallmadge, former governor of Wisconsin.

How long ago had his sister died? Fifty raps sounded, one for each year.

Had she died of illness? No answer.

"An accident?"

Yes.

"Was she killed by lightning? Was she shot? Did she fall from a carriage? Was she lost at sea?"

No answer. No answer. No answer. No answer.

"Was she thrown by a horse?"

Two knocks. Yes.

Definitely.

After he left, Cooper told his companions that every answer had been correct. He had been thinking about his sister, who, fifty years ago that month, had been killed when her horse threw her.

Cooper decided not to return. He was spooked. So was Greeley, but he described the visit anyway in his influential newspaper column. It would be the "basest cowardice," Greeley said, to deny that sensation of spirits knocking at the door.

Most mainstream church leaders loathed the spiritualist movement and condemned the Fox sisters, almost immediately, as a "nemesis of the pulpit." There was nothing of a biblical God, none of the teachings of Jesus Christ, in these tales from other side. The movement drew congregation members away from traditional teaching. Swedenborg's church reported a rush of new members; other "churches of spiritualism" flung their own doors wide.

The term *medium*—a person who provided a medium through which spirits could connect to the world of the living—became part of the common language. So did *sensitive,* meaning someone claiming an unusual sensitivity to messages from the summerland, the borderland, the spirit world, the seventh heaven, the misted realms where the dead wandered, waiting their chance to return. Professional mediums and fortune-tellers advertised in the local papers, hung their signs out to attract customers, offered sittings in their parlors. Newspapers that catered to the growing audience of believers—*Zoist, Light, Banner of Light*—claimed new subscribers every day.

In the early 1850s America, especially, seemed possessed. The spiritualist publications claimed that at least two million solid citizens could be counted as believers, perhaps half again that many in Europe. Many believed they had themselves talked to the dead. Not everyone shared the Fox sisters' talent for calling spirits out of the woodwork. But most people could manage the new craze of table tilting: gather a group around a pedestal table, place hands above the table's edges, fingers just touching the wood, watch the table bobble back and forth in response to questions asked, speculate about the power of the spirits to move material objects.

It was spiritual power at work, people said. Invitations to "tea and table-tilting" became standard social events. Others labelled their events table talking and invited professional mediums to join their parties. People said that when a gifted psychic joined in, tables did more than tilt and wobble. They hopped, crackled, hummed like a vibrating string. Some rose into the air, as if being tugged by invisible hands.

In 1853, three newly published books made the response of the clergy bitterly clear. The publications unanimously warned that table talking belonged to the devil. One author said he'd proved it through the power of the Bible. Merely placing the Good Book upon a levitating table, he said, would return it to the floor, subduing the evil spirits. Another writer titled his book *Table-talking: Disclosures of Satanic Wonders*.

A LETTER IN the *Times* of London, also published in 1853, signaled that other combatants were prepared to do battle with talking tables. The letter was signed by the physicist Michael Faraday, whose talent for invention had helped power the industrial revolution. His letter served notice that the science community was paying attention to the paranormal, but not in the way that Catherine Crowe had hoped.

Faraday was a prototype of the brilliant nineteenth-century scientist. He'd been revolutionizing science at the Royal Institution in London since 1808. He experimented with chemistry, showing how to liquefy chlorine (demonstrating that an element could transit from gas to liquid, a transition known as a phase change). He isolated the compound benzene, which became a critical component in motor fuels. Faraday followed these feats by developing an electric generator in 1831, a device he called a dynamo, and a prototype electric motor. He went on to design a simple battery and to experiment with creating a transformer. Many doubted that industrial production could have advanced so rapidly without him.

Faraday's letter to the *Times* concerned another experiment he had just completed that had nothing to do with invention at all. It was a laboratory test of a talking table.

For this particular study, Faraday took two flat pieces of wood and placed glass rollers between them. He fastened the device together with India rubber bands. With the rollers tucked in place, if a person pushed on the upper board, it would slide over the lower board. An instrument attached to the upper board was set to record even the smallest motion. Faraday then asked sitters to gather themselves around the table, fingertips resting on the edges of the upper board. It moved, despite the insistence of participants that they were sitting perfectly still. But there was no mystery to it, he declared, no spiritual magic.

As the instrument recorded, again and again, the people touching the board were pushing it sideways, sliding it along the rollers. The experiment showed that table tilters were often unaware of their own actions. As Faraday explained it, the board was responding to unconscious muscular twitches, "mere mechanical pressure exerted inadvertently by the turner."

In his letter to the *Times,* and in another to the *Athenaeum,* Faraday dismissed every theory put forward by the spirit believers. There was no ghostly energy guiding the tilts, no electrical force generated by dead-to-living communication, he said. He didn't want to hear any more pseudoscience from people who "know nothing of the laws" of electricity and magnetism. He'd been pestered enough by the superstitious, and his aim in writing the public letters was to restore some sanity to the discussion: "If spirit communications, not utterly worthless, should happen to start into activity, I will trust the spirits to find out for themselves how they can move my attention."

FARADAY'S SOBER call to sanity did not have much effect—at least not the effect he had hoped for—on a public caught in the thrall of spiritualism.

People seemed instead to be attracted to yet another medium, even more eerie than the Fox sisters—the impossible, unearthly Daniel Dunglas Home.

Born in Scotland in 1833, Home had emigrated from Edinburgh to New York as a child. Raised by an aunt and uncle, he grew up a soft-spoken child with a gentle, affectionate manner. When he was seventeen, his aunt threw him out of the house anyway. He'd become a child of the devil, she said. Tables floated when he entered a room, and he laughed when his frightened cousins shrank from an airborne chair. He wasn't safe to have around.

By the time he moved to England, Home swirled mystery around himself like a magician's cape. Tall and slim, so pale from underlying tuberculosis that he appeared near translucent in appearance, silver eyed and copper haired, he seemed a man made for magic. From American spiritualists came extraordinary tales of Home's feats: Knickknacks moved without being touched. Knocks sounded. Lights glimmered and faded like fireflies at dusk. Muffled voices whispered in empty corners. The wings of invisible birds rustled overhead. Ghostly hands touched people and then melted to mist.

"We were touched by the invisible," the poet Elizabeth Barrett Browning wrote in the summer of 1855, after a sitting with D. D. Home. She found the spectral hands completely believable and the medium completely wonderful. Her husband and acclaimed fellow poet, Robert Browning, dragged to the affair by his wife, resentfully admitted only to a peculiar evening. One hand had literally crawled, spiderlike, up Home's arm, Browning said. Another had tapped Browning on the shoulder, "a kind of soft and fleshy pat."

It was the table that really bothered Robert Browning. The table had risen off the floor at Home's bidding. The medium then invited Browning to inspect it. Browning felt around and under the legs. He watched Home's fine-boned hands, kept clear above the table, as the oaken legs shuddered into the air.

"I looked under the table and can aver that it was lifted from the ground, say a foot high more than once, with no wires or rods to be seen. I don't in the least pretend to explain how the table was uplifted all together."

Home gave his own explanation. If tables floated, if misted forms drifted across the room, if bells jangled on their own, each served only as evidence—the kind that science could never produce—of powers from beyond. He was a conduit, he explained, a messenger from the beyond.

"I believe in my heart that this power is being spread more and more every day to draw us nearer to God. You ask if it makes us purer? My only answer is that we are but mortals and as such liable to err; but it does teach that the pure in heart shall see God. It teaches us that He is love and that there is no death."

He promised, as Swedenborg had before him, that he could help people see God, prove to them that the afterlife could be seen and touched. "Fear not," wrote a ghostly hand to a young woman who had frozen into a white silence during one of his séances. And as she shrank back, a heavy bookcase—"one that would at least require four men to move"—began to trudge ponderously toward her, thumping across the floor.

In Home's soft voice, in his words of faith and hope, even a walking bookcase could seem, somehow, shining with the dust of angels. "Fear not, Susan, trust in God. Your father is near. He is the Great Father."

IN THE SPRING OF 1857, a tired and cranky team of scientists made its way from Harvard University to upstate New York, determined to try yet again to rub some of the superstition out of modern culture.

They'd been sent to investigate claims that two sons of a Buffalo police officer were able to summon spirits into a theatrical performance. There was no talk of God and holy messengers from Ira and William Davenport. Instead there were rolling hoops, ringing bells, twanging mandolins brought to insane life while the mediums sat tied to their chairs. But following Home's lead, the Davenport brothers gave full credit to the spirits for enlivening their events.

Such spirit shows now drew audiences in villages and cities across the land. There were so many, and so many were so obviously phony, that the editor of the *Boston Courier* offered $500 to any medium who could really produce spirit phenomena. His one condition was the results had to be verified by Harvard University. The university administrators preferred to take the higher road of scoffing at such productions. But they also believed that if some reputable professors took on the job, spiritualism could be easily exposed and, they hoped, eliminated. Over the protests of the designated investigators, Harvard's president sent his professors to Buffalo.

The Davenport Brothers—as they billed themselves—had pioneered a technique for inviting spirit participation, a large wooden box that they called a cabinet. The enclosed space, they claimed, was necessary to "condense the psychic energy." Of course, no one could see the Davenports while they were inside, something that other traveling mediums noticed, admired, and imitated.

The Davenports' cabinet looked like a wide walnut armoire, with three doors in the front. The center door held an oval window, curtained by dark velvet. Behind it were two wooden chairs, placed so that when the brothers sat down, they faced each other. Members of the audience could enter and tie the Davenports' hands and feet. They could bind Ira and William together if they wanted. They could check the knots, seal the ropes with wax. Once the satisfied audience sat back down, once the doors closed, the cabinet slowly came to exotic life. The bolts on the doors mysteriously slid into place, phantom hands appeared and retreated through the velvet curtain, musical instruments placed inside played familiar tunes. (This was twenty years before Edison introduced recorded music.) When the watchers reopened the cabinet, the brothers still sat tied to their chairs, unbroken wax sealing the knots.

The Harvard investigators required more than a few cords and some dabs of wax. They brought a wagonload of rope. Local newspapers estimated the scientists had five hundred feet of good hemp cord with which to tie the brothers' hands and feet. The professors also bored holes in the cabinet and ran the ropes out through the openings, back in, out again. They knotted the cords, again and again, into a net around the outside walls. Amused reporters noted that the observers could barely make out the cabinet through Harvard's skein of rope.

Still, as journalists wrote, the mandolins yet twanged and the spirit hands still fluttered. They also wrote that the university refused to publish a full account of the day with the Davenports. The professors issued a report condemning the performance as a trick, but added that the findings didn't need to be dignified with details.

The newspapers thought that they did, actually. And the townsfolk whispered that one academic left the cabinet wound with his own rope like a fly trapped by a vindictive spider. Harvard denied that rumor to the Boston papers, but in Buffalo, folks thought the professor lied.

The Davenports may have been charlatans, but they were charming and capable ones. People liked them—to the point, perhaps, that they even trusted the performers' explanations of spirit mischief. Aloof and autocratic university scientists with their rational cynicism seemed far less appealing.

MANY YEARS LATER, William James marveled at the ineffectiveness of such scientific strikes against the supernatural. "How often has 'Science' killed off all spook philosophy, laid ghosts and raps and 'telepathy' away underground as so much popular delusion?" he would wonder ironically. As James noted, the ghosts kept coming back, the visions yet glimmered, the voices yet sounded. No matter how many times scientists evoked mental illness, dreams, fantasy, and stupidity as explanations for bumps in the night, people kept reporting them as though they were real.

Go back in history, James pointed out, and "you will find there was never a time when these things were not reported just as abundantly as they are today." Ghosts drifted like smoke through the pyramids of Egypt. Smoldering demons climbed out of fires in ancient Africa. Spirits walked with native hunters in the American forests, guiding the arrows with their invisible hands.

"The phenomena are there," he wrote, "lying broadcast over the surface of history." The question for him had never been whether people saw—or thought they saw—ghosts. The question had always been what to do with such odd reports, how to classify such irregular events, where to place them in our orderly descriptions of how the world worked. Or maybe it was better described as a problem, one that didn't fit in such organized systems.

"The ideal of every science is that of a closed and completed system of truth," James acknowledged. If supernatural events did not match the categories of the scientific system they "must be held untrue." James admired the efficiency of the "scientific" approach to the spiritual murkiness. "It is far better tactics, if you wish to get rid of mystery, to brand the narratives themselves as unworthy of trust," James wrote. But while he agreed that most so-called supernatural events were suspect, he worried that scientists stayed deliberately blind to the rare credible ones, that researchers might be ignoring "a natural kind of fact of which we do not yet know the full extent."

And he worried too about the larger effect of such prejudice on the way people viewed science itself. "Thousands of sensitive organizations in the United States today live as steadily in the light of these experiences, and are as indifferent to modern science, as if they lived in Bohemia in the twelfth century. They are indifferent to science, because science is so callously indifferent to their experiences."

If scientists did not afford some respect to the beliefs of the lay public, James warned, there was little reason for the public to respect the pronouncements of science.

BY THE TIME William James was thirteen years old, he had attended almost a dozen different New York schools. By 1857, the year Harvard took on the Davenports, the James family had been back in Europe for two years. Their father had decided on the more sophisticated, "sensuous" advantages of a European education. He took the family to Switzerland, but didn't like the schools there either. Then London. Then France, where they moved five times in the next two years. Then back across the Atlantic to Newport, Rhode Island. They stayed three months before moving back to Switzerland. There William studied painting, and Henry was sent to a polytechnic school. Then Germany for a year. Then back to Newport.

It was gypsy traveling and scattershot education, and every one of the James children considered it destructive. Henry James Jr. deliberately omitted two of the moves from his adult autobiography to make the family appear more normal. Alice grew to be an unstable and bitter woman. "If I had had any education," she wondered once, "would I be more or less of a fool than I am?"

In 1861 William enrolled at Harvard. The university would serve as his first real anchor, mooring him against the erratic winds of home life and the wilder storm blowing across the country, the bloody howl of the Civil War. He'd intended to become a soldier, signing up at the age of nineteen as a volunteer with the Newport Artillery Company, making himself available for state militia recruitment. Almost until that day, Henry Sr. had refused to consider allowing his eldest son to leave home for university. Now he told William to start immediately. He wanted Henry Jr. there as well. His two younger sons he encouraged off to war the following year, making it obvious, as he would throughout their lives, that he treasured his older sons far more. "I've had a firm grip upon the coat tails of my Willy and Harry, who both vituperate me beyond measure because I won't let them go" to fight, James Sr. wrote to a friend.

The war—and their father's response to it—would leave a legacy of bitterness in the James family, echoing the lingering damage across the country. The older brothers long harbored guilt, and the younger ones a sense of rejection. It was years later, toward the end of the nineteenth century, before it was clear that William James had made his peace with himself and his kin. In a speech honoring the Civil War dead, he praised the individuals, the soldiers like his brothers, but not the war itself. Nations, James said simply, are not saved by wars. They are saved, he said, by "acts without external picturesqueness; by speaking, writing, voting reasonably; by smiting corruption swiftly; by good temper between parties—by people knowing true men when they see them, preferring them as leaders to rabid partisans and empty quacks."

He spoke in memory of his brother Wilkie, dead at the age of thirty-eight, largely due to the long-term crippling damage of battle wounds. But he spoke for himself too. Rabid partisans and empty quacks? William James could never abide people who pretended that there could be an honorable stand without honesty behind it.

> *Now, don't, sir! Don't expose me! Just this once!*
> *This was the first and only time, I'll swear,—*
> *Look at me,—see, I kneel,—the only time,*
> *I swear, I ever cheated . . .*

The year was 1864, the Civil War yet burned in the United States, and William James was in his third year at Harvard when Robert Browning published that bitter portrait of the archetypal professional medium.

Browning's wife, Elizabeth, had died three years earlier, after years of illness. He admitted that he would not have published the poem during her lifetime. She was too ardent a believer, laughingly describing herself once as a follower of every goblin story. "Smile," she wrote to her sister, "but such things are so indeed."

Browning indulged her, his lovely, hopeful, delicate wife, with whom he had eloped to Italy in 1848. Theirs was one of society's more famous romances—the young Robert Browning courting the acclaimed poet Elizabeth Barrett. Gossips whispered of her possessive father's fury, their flight in the middle of the night, their romantic idyll in Italy. The marriage lasted thirteen years, until her death in 1861. She wrote some of her best poetry when they were together, including *Sonnets from the Portuguese,* with all its soaring romanticism and famous lines: "How do I love thee? Let me count the ways."

Browning himself wrote only sporadically during the marriage. After her death, though, he threw himself into a series of intense and emotional story-poems, collected in 1864 into a book called *Dramatis Personae.* Among them, the poem "Mr. Sludge, the Medium" stood out like a thistle in a cluster of red velvet roses, a spiky blast of rage in the midst of love and longing and loss.

In Browning's poem, Mr. Sludge first begs to have his mediumistic tricks kept secret and then defiantly confesses anyway: he hired children to listen at the keyhole and pass along information about clients. He made ghost hands appear above the table by fixing padded gloves onto a rod attached to his shoes. He used phosphorus to make spirit lights glow during his séances.

To turn, shove, tilt a table, crack your joints,
Manage your feet, dispose your hands aright,
Work wires that twitch the curtains, play the glove
At end o' your slipper,—then put out the lights.

As the medium trade grew, its flaws and its frauds had become more obvious. The Davenport Brothers, after their popular triumph over Harvard academics, had brought their cabinet show to England, only to be caught in some obvious chicanery with ropes. In Liverpool, an angry audience had smashed their cabinet to pieces, later selling the splinters as souvenirs. Ira Davenport called it "a nauseating example to all foreigners of 'ow the average Englishman does things at 'ome", and the brothers took their act to the Continent.

But later Ira would reveal many of their better rope tricks to the magician Harry Houdini. The fact was, too many actual Mr. Sludges existed, and too many of them shared their secrets. Professional magicians and actors took to the stage, performing to packed houses, showing that they too could conjure like the Davenports, or, as one theater poster declared, could reproduce "All the Tricks of the Spirit Conjurers."

One female medium was found to have an ingenious wire dummy, covered with a thin skin of rubber, which could be inflated during a dark séance to resemble the "spirit" form of a small child. Deflated, it could be folded and worn as a stylish bustle, neatly concealed in her skirts. Others hid thin packets of clothing—preferably cobweb-fine French muslin—in their undergarments. The cloth was made bright in spots by luminous oil, made of phosphorus and ether. The oil glowed only faintly in the dark, but bits of glass or paste diamond sewn into the cloth added a brighter glitter. Not surprisingly, spiritualist performers preferred to hold their séances in the dark and in their own homes. In one medium's abode, investigators found a trapdoor under a cabinet, opening into a passage that led into a backroom that contained another trapdoor out of the building itself.

Such exposés delighted the critics of the spiritualist movements. Scientists rejoiced in what they hoped would lead to a new skepticism. And religious leaders, dismayed to find spiritualist churches the fastest-growing houses of worship, renewed their efforts to discredit the movement. In a book called *Spiritualism Unveiled,* a prominent theologian described spiritualists as wife beaters, sexual deviants, and anti-Christian: "They are as unlike in their moral influence as are Christ and Belial."

For the moment, it was the latter accusation that most irked the spiritualists, and they were quick to hit back with their own perspective on the morality of the day. "Are those who play tricks and fling about instruments spirits from Heaven?" inquired one spiritualist newspaper in an impassioned editorial. "Yes, God sends them, to teach us this, if nothing more: that He has servants of all grades and tastes ready to do all kinds of work and He has here sent what you call low and harlequin spirits to a low and very sensual age."

· · ·

BUT ROBERT BROWNING wasn't writing only about the Davenports and
their ilk when he penned "Mr. Sludge." As everyone knew—at least, everyone in literary circles—the talented, amoral, fluently devious Mr. Sludge
was a barely disguised caricature of Daniel Dunglas Home.

Home had returned to England after a sweep across Europe so successful that the Roman Catholic Church charged him with witchcraft. While
in Russia, he had married a young goddaughter of the tsar with due pomp
(the French novelist Alexandre Dumas, author of *The Three Musketeers,*
stood as best man, Alexei Tolstoy as groomsman). Respectability had done
nothing to alter Home's ethereal, inexplicable nature. He was "Stranger
than Fiction," according to a magazine article about him written in 1860,
still widely considered the best description known of the medium—and
the best counter to Browning's long-standing dislike of the man. The magazine writer described a sitting with Home in which the medium "rose
from his chair, four or five feet from the ground. . . . We saw his figure pass
from one side of the window to the other, feet foremost, lying horizontally
in the air. He spoke to us as he passed, and told us that he would then
return the reverse way and re-cross the window, which he did."

This description appeared in a new British literary magazine, *The
Cornhill,* edited by the famously sharp-tongued writer William Makepeace
Thackeray, whose 1848 novel *Vanity Fair* had established him as a man
more than willing to skewer pretensions. Yet Thackeray not only approved
the article, he wrote an editor's note vouching for the "good faith and honorable character" of its author. Thackeray had friends in scientific circles.
He had no doubt they were unhappy; he found himself accused, outright,
of being complicit in the spread of spirit fraud. When confronted by one
furious scientist, Thackeray simply cited his own experience: "It is all very
well for you, who have probably never seen any spiritual manifestations to
talk as you do; but had you seen what I have witnessed, you would hold a
different opinion."

To his believers, Home offered something that neither irate scientists
nor resentful poets seemed able to counter. He boasted a flamboyant mystic
appeal that the staunch mainline Protestant churches of the establishment of
America and Britain—Anglicans, Episcopalians, and Methodists—neither

could nor would attempt to match. Perhaps most powerfully of all, he and his counterparts offered a glimpse of possibilities through open doors that their critics—churchmen, skeptics, and scientists alike—would slam without a glance.

This was the reality that William James, scientist and scholar, saw as he made his way from his father's shuddering haunted walls to the more steadfast halls of academia. It was the reality that he looked out upon and wondered about. Even as he built himself a respected place among the door-slammers of science, James began to think, more and more, that it was unconscionable not to look out through that portal, to see what might possibly be there.

It was past time, he thought, for science to open its mind.

2

*

A SPIRIT OF UNBELIEF

*W*HEN CHARLES DARWIN published his famous (or infamous) book *On the Origin of Species* in 1859, the tensions between Victorian science and religion became undeniable, obvious for all to see.

The Darwinian idea of evolution, that species take shape slowly—ruthlessly sculpted by natural selection, random mutations, and environmental pressures—was a difficult proposition for most Victorians to embrace. Even scientists hesitated at first to choose "survival of the fittest," as it came to be called, over the Genesis creation story. There were geologists who flatly refused to accept what the fossils they found said about the age of Earth; science textbooks of the time routinely acknowledged a Creator. Mid-nineteenth-century British society was, at least on the surface, invested in a scriptural version of how the world came to be. After all, the queen was also official head of the Anglican Church.

The few people who suggested otherwise, before Darwin brought the argument fully into the open, tended to publish their works anonymously. One author who proposed that the universe might have developed according to natural laws, without divine intervention, insisted that his name remain anonymous until his death. Even so, the bishop of Oxford used his book as the foundation for a sermon condemning mid-nineteenth-century science and what he called its "mocking spirit of unbelief," adding that the author, whoever he might be, was obviously no gentleman.

Everyone—even those who no longer insisted on a literal reading of biblical accounts of creation—could see an either/or dilemma taking shape. Choose science or faith; choose the Bible or *Origin*. Rightly or wrongly, there came a perception of a widening rift between incompatible, mutually exclusive realities.

In 1858, both Darwin and his fellow naturalist Alfred Russel Wallace were presented to the British scientific community as coauthors of the theory of natural selection.

Coauthors, yes. Coequals, no.

Darwin was lead author. He'd been developing and documenting his theory for more than twenty years, testing it on friends, anticipating hostility, waiting to reveal it until he could counter every potential argument. His confidants had feared that Darwin would never quite be done, that he would polish his points into the infinite future. So Wallace was the catalyst, the spark, the innovative thinker whose independent realization of natural selection flushed the wary Darwin out of hiding.

It spoke much for the theory that two such different minds could independently realize its power. Darwin turned fifty in the year he published *On the Origin of Species*. His background was affluent, upper class. Cambridge educated, he had served as naturalist on a sea voyage to South America and around the world. With the natural treasures he collected at stops such as the Galapagos Islands, he had returned to his English country home to classify barnacles, ponder the adaptive variety of beaks among the finches of the Galapagos, and painstakingly document his findings as he worked out his daring theory of evolution. Darwin's nature—reclusive and self-critical—made him reluctant to engage in a public fight. Until the moment he received Wallace's paper, Darwin was still debating the right moment to make his stand.

Wallace was thirty-five when he was declared a coauthor of the theory of evolution. He was tall, thin, energetic, with bright blue eyes framed by wire-rimmed glasses. From the middle class and largely self-educated,

Wallace had become intrigued by plant species and their habitats while working in his brother's surveying business. Leaving that work, he helped to organize two collecting expeditions, one to South America and the second to Malaysia, then known as the Malay Archipelago. His travels kept him mostly out of England between 1848 and 1862. Wallace's paper arguing for natural selection was read in England while he was still in Malaysia, assembling a collection of more than 125,000 specimens.

Darwin and Wallace were both enormously influenced by their foreign expeditions, which Darwin illuminated in his earlier book *The Voyage of the HMS Beagle.* Far from England's manicured landscapes, they explored places that displayed a wonderfully untamed diversity of life. Each man was struck by the way that species seemed to have adapted—and were still adapting—to environmental pressures. Wallace had been in Malaysia for five years, tracking the shifting patterns of plants, birds, and insects, when he first wrote to Darwin, sharing his early ideas on how species might respond to changes in location. A year later, in 1858, Wallace sent Darwin a copy of a paper, "On the Tendency of Varieties to Depart Indefinitely from the Original Type," which detailed his new theory of natural selection.

The paper neatly outlined major ideas that Darwin had been compiling into his own book. While recuperating from a tropical fever, Wallace had decided to focus on one particular question: In both the human and animal worlds, why do some die and some live? His observations told him that success seemed to foster success. The healthiest tended to stay healthy. "The strongest, swiftest and most cunning" escaped their enemies. The most able hunters avoided starvation. "Then it suddenly flashed upon me that this self-acting process would necessarily *improve the race,* because in every generation the inferior would inevitably be killed off and the superior would remain—that is, *the fittest would survive.*"

One read through Wallace's manuscript told Darwin that he had to move forward or lose claim to his cherished theory. A year later, Wallace's paper was jointly presented with Darwin's at a London scientific meeting. The year after that, in November 1859, *On the Origin of Species* went on sale at a price of 15 shillings. It promptly sold out its print run.

Darwin's genius went far beyond his ability to make a reasoned and researched case. He possessed a gift for combining science with everyday, commonsense observations that could be shared by the farmer, the gardener, and the recreational hiker. Geology reveals our history, Darwin said, and fossils tell of species come and gone. But we can see selection at work *now*, in the successful breeding of garden flowers and farmyard animals, in the natural variations in life around us in our fields and forests, in everything showcasing the incredible, responsive, ever-changing diversity of life.

From the planet's simple beginning, Darwin wrote, "endless forms most beautiful and most wonderful are being evolved." The evidence is sprawled before humankind: from the traces of dead creatures, cradled in the rock beneath our feet, to the shape-shifting existence of birds and of butterflies brightening the air around us.

"It is so easy," he added, "to hide our ignorance under such expressions as 'plan of creation' & 'unity of design.'" Darwin feared his own generation would never get past its biblical baggage. He thought it might be too difficult for his contemporaries to abandon the idea that Earth was young, that species arose finished in their nature, that humans stood separate from all else, that a divine intelligence had shaped life to meet its particular standards.

He saw that reluctance and anxiety entrenched among fellow scientists. "I look with confidence to the future," Darwin wrote, "to the young and rising naturalists, who could face the realities of life without prejudice." Darwin was recommended to Queen Victoria as a candidate for knighthood the month after his book was published. But he never became Sir Charles. The bishops of Her Majesty's Anglican Church made sure of that.

PERHAPS THE MOST famous of the ensuing debates between a supporter of evolution (a scientist) and a critic (a member of the clergy) took place in 1860.

The debaters were Samuel Wilberforce, the bishop of Oxford—who had been instrumental in preventing Darwin's knighthood—and T. H. Huxley, physician and scientific scholar. This was the same Huxley who would coin the word *agnostic* to describe himself and his belief that God, or the ultimate reality, is unknowable.

The bishop decided to tackle evolution on scientific rather than theological grounds. He argued that species were ever fixed, permanent in their shape, "a fact confirmed by all observation" of early man and his domestic animals, such as could be found in the tombs and pyramids of the ancient Egyptians. "The line between man and the lower animals was distinct," the bishop continued. "There was no tendency on the part of lower animals to become the self-conscious, intelligent being, man, or in man to degenerate" in the direction of lesser species. He then, according to published accounts, asked Huxley whether it was through his grandfather or his grandmother that he claimed descent from a monkey.

By nineteenth-century standards of debate, this was an outrageous display of rudeness. It elevated the exchange to a level of near mythological proportions, told and retold. As recounted in *Macmillan's Magazine,* "On this remark, Mr. Huxley slowly and deliberately arose." Huxley stood there, thin and pale, quiet and very grave, the magazine reported, and replied: "He was not ashamed to have a monkey for his ancestor; but he would be ashamed to be connected with a man who used great gifts to obscure the truth."

The debate helped make Huxley almost as well-known as Darwin, and he took advantage of it in further debates. His ferocity and tenacity in defense of evolutionary theory earned him the nickname "Darwin's bulldog." Many thought science had triumphed, at least in this instance, over religion. For many others in England and America, however, the painful conflict only intensified a resolve to cling to deeply held beliefs that naturalists with their compiled data—and their clever comebacks—could never hope to explain.

ALFRED RUSSEL WALLACE returned to England in 1862 and strode without hesitation into this fray. He lacked Huxley's ability to turn a wicked phrase but he brought his own gifts: energy, enthusiasm, and sincerity.

Wallace traveled the country in support of Darwin and natural selection. He fearlessly affirmed that humans shared common ancestry with other animals, that our species was as easily explained by adaptation, by the selection of survival traits, as any other. Wallace had lived among the tribes along the Malay Archipelago. He assured his audiences—playing to their Victorian sense of self-superiority—that primitive societies represented humans in an earlier stage of development, less evolved than technologically advanced westerners.

While Wallace camped in a simple hut, foraging for food in the tropical forests, modern society had continued its industrial advances. The mass-produced paper bag, the photographic slide, the safety elevator, and the machine gun were all recent inventions. The blazing, impossible speed of light had been measured, emphasizing that even the golden aura of a late afternoon was a matter of physics, a calculation to be mastered by man.

But as he traversed England, Wallace gradually perceived dark spots in this polished progress. It seemed to him that the moral evolution of Western society did not match its intellectual development. He could easily enumerate examples. The slums of London stank with raw sewage; brothels catered to the deviant (some specialized in "birching," or whipping their customers); uneducated children stole for fun as well as for need. After more than a decade away, Wallace found his homeland version of civilization more violent, less compassionate, less decent, than that of supposedly "less evolved" tribal societies. "The mass of our populations have not advanced beyond the savage code of morals, and have in many cases sunk below it," he complained to a friend. It was possible, Wallace thought, that science was precipitating a loss of faith. And it was also possible that a faithless society might find itself in a state of backward evolution. It might be that without a God—or at least the belief in one—there could be no reinforcement of right and wrong, no bracing assurance of punishment and reward.

Although he found organized Christianity's way of explaining the world to be antiquated and unconvincing, Wallace began to reconsider the possibility of a moral force at work in the universe. He worried that if science denied even the possibility of such a higher power, the result could be a widespread amorality that would rip the social fabric. Wallace began to think that he and his colleagues bore a responsibility that they had thus far shirked. He became compelled by the idea that it was the duty of scientists to study not only the "physical parts of our nature, but the moral ones."

Uplifted by this new and growing sense of purpose, Wallace attended his first séance in 1865. As he would explain, Wallace thought of this as a scientific expedition into the dark jungles of spirit phenomena, worth the risk of giving ammunition to critics eager to discredit him. Pondering a dizzyingly radical new theory, he thought he might find the way to an integration of science with spirit.

Wallace's new idea was that natural selection had its limits, at least with regard to human beings. It could account for the physical body, yes, for skin, hair, muscles, the thump of the heart, flex of the lungs, shape of the hands, curve of the spine. These, he continued to believe, all evolved according to Darwinian (or Wallacean) principles.

But the mind, he proposed, was different. Perhaps intelligence, morality, that ephemeral thing called the human soul, developed along other lines. Perhaps our better nature was crafted by direction, by a power yet to be discovered; perhaps the design of the universe was such as to encourage spiritual development. Perhaps, Wallace proposed, even "the material imperfections of our globe" were not random at all, but purposeful, planned by a higher power. Perhaps "the wintry blasts and summer heats, the volcano, the whirlwind and the flood, the barren desert and the gloomy forest, have each served as *stimuli* to develop and strengthen man's intellectual nature; while the oppression and wrong, the ignorance and crime, the misery and pain, that always and everywhere pervade the world, have been the means of exercising and strengthening the higher sentiments of justice, mercy, charity, and love, which we all feel to be our best and noblest characteristics, and which it is hardly possible to conceive could have been developed by any other means."

It occurred to Wallace that evidence for such an artful planner could only be found by investigating the supernatural realms. As he saw it, his first move should be a feasibility study, an exploration into whether evidence could be gathered at all. He needed to know, for instance, if one could reasonably expect to gather information about spiritual powers. In his first sittings with London mediums, Wallace saw nothing that approached the level of scientific proof. But the séances were just weird enough to be encouraging. If nothing else, he could argue that he'd seen inexplicable things happen, things that had not—and perhaps could not—be explained by the laws of science.

In his notes, Wallace said he was particularly impressed by one table-tilting demonstration in which "a curious vibratory motion of the table commenced, almost like the shivering of a living animal. I could feel it up to my elbows." He was several times startled by the information provided by mediums. For example, a medium spelled out the names of a visitor's deceased relatives, backward and forward, even though the visitor had arrived at the séance anonymously. The cleverness of the spelling seemed to Wallace to be evidence of survival of intelligence after death. And, like so many before him, he found a séance with Daniel Dunglas Home particularly unsettling. A few of the phenomena "give me a solid basis of fact," he concluded, urging his fellow scientists to continue with him in this inquiry. After all, Wallace said, other intelligent men must be troubled, as he was, by mysteries "which science ignored because it could not explain."

As Charles Darwin promptly warned him, Wallace was sending the wrong message to their critics and lending unwarranted credibility to the concept of spirit powers. Darwin feared that Wallace now gave the impression that one of evolution theory's founders had abandoned science in favor of superstition.

"You write like a metamorphosed (in retrograde direction) naturalist," Darwin wrote furiously. "I *defy* you to upset your own doctrine." In his outrage, though, Darwin missed a crucial point. Alfred Russel Wallace had not and never would turn away from the theory of evolution. He promoted it and worked to refine it all of his life, even into the twenteeth century, long past the time Darwin—who died in 1882—was around to scold him.

It wasn't that Wallace rejected his theory. It was that he found it less than satisfying. Basic survival and mechanical evolution, he decided, were not enough.

"I FEEL CONVINCED that English religious society is going through a great crisis now," wrote a Cambridge University lecturer in 1867. "And it will probably become impossible soon to conceal from anybody the extent to which rationalist views are held, and the extent of their deviation from traditional [Christian] opinions."

The writer was Henry Sidgwick, a respected member of the classics faculty at Trinity College, Cambridge. Within the following decade he would publish his book *Methods of Ethics,* hailed as a major work of moral philosophy in the tradition of John Stuart Mill and Immanuel Kant. And in 1882 he would found, along with two friends, Frederic Myers and Edmund Gurney, the British Society for Psychical Research. Gurney would in turn recruit William James into their cause, as an extension of an easy friendship between the two men.

The movement—some would call it a quest—began first in England, fomented by Wallace, stirred by the kind of hostile debates staged by Huxley and Bishop Wilberforce, sought out by those who craved a refuge from the increasingly belligerent stands taken by both religious and scientific leaders. In an era when Darwinians faced off against the defenders of Genesis—neither side allowing for a middle ground—both groups lost a measure of credibility and trust. The psychical research movement rose in response to such rigidity, built by those who believed that objective and intelligent investigation could provide answers to the troubling metaphysical questions of the time—and that those answers mattered.

The son of a clergyman, Sidgwick had reluctantly abandoned Christianity as a system unable to keep up with the present. "God owns the past," he told a friend, but not the present. Yet, like Wallace, he worried about humankind stripped of faith. Without a religion—without a deity promising punishment and reward—Sidgwick wondered what would bind people to principles of honor and decency.

The scholar Sidgwick pondered great cultures that had relied on religion to set moral standards. The central questions of identity, of how to live and behave in the world, of how to right wrongs and avoid disaster, had traditionally been put to the gods—animal-headed Egyptian gods and the pantheon of the Greeks and Romans. Life's big questions were laid before Allah and his Prophet. They were addressed in Arab mosques, Buddhist monasteries, Jewish temples, and European churches. Sidgwick wondered where people would turn if they accepted that the only source of Truth, with a capital *T,* came from modern science, that the only answer was that life arose from random, mechanical, materialistic forces, that it was governed by none but physical principles. He shuddered at the empty silence of what he called "the non-moral universe."

In personality, Sidgwick seemed an unlikely candidate to rally others to a cause. Slight, with a thin face and wide gray eyes, he had a shy smile and a hesitant way of speaking, made more so by a faltering stammer. But he was respected on campus as an unusually fair-minded man, and a decent one. Under the intellectual reserve, he was compelled by a desire to overcome his personal failings and by a steadfast determination to make a difference.

"When I found out how selfish I was," Sidgwick wrote to his sister, shortly after he moved to Cambridge, "I used at first to try and alter myself by conscientious struggles, efforts of Will." He made a "golden rule" for himself not to think about himself more than half an hour out of every twenty-four. He deliberately hunted for good causes without enough support, making a priority of women's right to education and everyone's right to seek answers to questions they cared about, even those dismissed as nonsensical or the stuff of superstition.

His cousin Edward White Benson (later archbishop of Canterbury) first enlisted Sidgwick in the latter cause. Having helped found a "Ghost Society" at Cambridge, the outgoing Benson drafted his quiet cousin to visit some local mediums and psychics. From the first, Sidgwick approached the subject with characteristic tough-mindedness, easily detecting the use of mechanical devices and sleight of hand, writing to his sister, "I gained nothing but experience in the lower forms of human nature."

But the idea of being able to prove that there was something more, a spirit existence, a power beyond that of human grasp, intrigued him. And as he continued to investigate, he thought he glimpsed, only occasionally, a glimmering spark of something unexpected. In another letter to his family he wrote of his sense of grasping at handfuls of smoke while somewhere within the billows burned a genuine flame.

One of Sidgwick's students, Frederic Myers, was quick to see the real purpose in his forays into the occult—and to follow. Born in 1843, Myers was, like Sidgwick, the son of a well-to-do Yorkshire clergyman and an unusually clever child. Myers expressed his first doubts about his qualifications for heaven at the age of two, wrote his first sermon at five, and entered Cambridge when he was seventeen, still fired with faith, praying to be stronger, wiser, to "have a strength not my own infused into me."

Yet the more Myers studied, the more he learned of science and history, the more heaven seemed to slip from his grasp—or his sense of reality. The Anglican Christianity of Yorkshire and Cambridge began to look frail and dusty. It seemed to him more suited to the static past than to the dynamic present. Darwinian science troubled Myers, but it troubled him more that the church was so resistant to new ideas, even ones that might improve lives.

The cause of women's education first brought Myers and Sidgwick together. The Anglican Church insisted that God intended women to be subservient to men. So did most of society. Their own university, as most others, barred women from obtaining degrees. Queen Victoria herself was adamant on the subject: "Were women to 'unsex' themselves by claiming equality with men, they would become the most hateful, heathen and disgusting of beings and would surely perish without male protection."

Most of Sidgwick's friends, his peers at Cambridge, saw only trouble in his argument that women should be allowed the "immense educational influence" of training for a profession. Sidgwick called it a matter of "simple justice." Such "justice," he was warned in turn, would lead to further demands by women, further concessions by men.

Myers, on the other hand, agreed with Sidgwick without hesitation—so readily that it caught the professor by surprise. He had known Myers largely as one of his brighter students—a tall, gray-eyed young man with carefully elegant clothes and smoothly waved brown hair. Sidgwick was pleased to find behind the stylish exterior a classics scholar with the heart of a crusader.

It was natural that their conversation should drift to Sidgwick's other unpopular cause. In his spare time, Sidgwick continued to investigate for the Ghost Society, sifting for evidence of spirit communication. Myers was fascinated by what occurred in those darkened rooms; in their faint hints of a life after death, he saw a way to try to resolve the questions and doubts that had been so far intractable. In those early supernatural investigations, Myers saw the shape of a bigger quest, "an endeavor to learn the actual truth as to the destiny of man."

Years later, describing his part in building an organization dedicated to psychical research, Myers put his role in self-deprecating perspective. He was the dreamer of the group: "Edmund Gurney worked at the task with more conscientious energy; the Sidgwicks [Henry and, later, his wife, Nora] with more unselfish wisdom; but no one more unreservedly than myself . . . staked his all upon that distant and growing hope."

FOLLOWING HIS SPIRITUAL insights, Alfred Russel Wallace in the late 1860s invited a series of respected scientists—including a noted physiologist and rather brilliant physicist named John Tyndall—to attend private séances at his home and investigate the phenomena. As he explained to another invitee, T. H. Huxley, Wallace hoped that his colleagues would consider such study "a new branch of Anthropology."

As might be expected, Huxley's response was more barbed than the other rejections. Huxley had been to a few séances earlier in his life, and he'd found them ridiculous. Wallace might be onto something, but Huxley wasn't impressed by whatever it was: "It may all be true, for anything I know to the contrary," he wrote in reply to the invitation, "but really I cannot get up any interest in the subject."

Undaunted, Wallace pressed him again; many of the scientists had flatly refused, and Tyndall had stayed but a few minutes before stomping out the door. Wallace knew Huxley from their mutual labors in favor of natural selection; he hoped that collegiality would persuade him. Huxley's second refusal was more pointed. As he explained, he'd heard enough of spirit communications to know that they were so much nonsense. He didn't need to bore himself senseless by sitting through a séance in person: "The only good argument I can see in a demonstration of the truth of 'Spiritualism' is to furnish an additional argument against suicide. Better to live a crossing-sweeper than die and be made to talk twaddle by a medium hired at a guinea a séance."

Finally, Wallace accepted that his circle of acquaintances seemed more likely to ignore his new science than advance it. He decided to give his ideas a more public airing, and in April 1869 published a paper laying out his view that natural selection might have limits, that an "overruling intelligence" might be responsible for the development of mental and moral behavior.

Wallace undoubtedly felt driven to be combative. He complained that a scientist seeking to explore the supernatural found himself instantly demoted, "set down as credulous and superstitious, if not openly accused of falsehood and imposture, and his careful and oft-repeated experiments ignored as not worth a moment's consideration." But it wasn't just exasperation that prompted Wallace to air his argument. It was his sense of rightness, his conviction that understanding supernatural events could help illuminate "the nature of life and intellect, on which physical science throws a very feeble and uncertain light." Wallace believed, he said, that all branches of science would suffer until such inexplicable happenings were seriously investigated and "dealt with as constituting an essential portion of the phenomena of human nature."

His colleague's righteous tone further alarmed Darwin. After reading an early version of the manuscript, he wrote to Wallace, "If you had not told me, I should have thought they [the comments] had been written by somebody else. As you expected, I differ grievously from you and I am very sorry for it."

But Wallace wasn't sorry. He had failed to win over Darwin or his illustrious group of allies, but he was moving on to better prospects. At long last, Wallace had persuaded another of Britain's more acclaimed scientists—the chemist and inventor William Crookes—to conduct a serious investigation of the ever-elusive D. D. Home.

WILLIAM CROOKES was a thirty-nine-year-old Londoner, a big man with narrow blue eyes in a narrow, high-cheekboned face, a dark beard, and a splendidly imperious manner. A designer of scientific equipment and a gifted chemist, Crookes had recently discovered a new element, which he named thallium. A soft, malleable metal, thallium was also a neurotoxin so potent that later generations of scientists would speculate that Crookes became involved in spirit research only because of work-related brain damage.

As Crookes told it, though, he hadn't found Wallace's arguments particularly convincing. He thought he might be able to help straighten out the errant naturalist. Crookes's initial plan was to scrutinize a few so-called mediums and finish the subject off.

One of the first mediums that Crookes visited was a thin, intense, small woman who held sittings in her neat little parlor. She seemed simple enough in technique, using one of the more popular devices of the time for spirit communication, a planchette.

A planchette was a heart-shaped piece of wood, mounted on small wheels so that it rolled. At its narrow end it held a pencil, point down so that it just brushed a piece of paper placed underneath. To operate a planchette, users put a hand on the wood, concentrated, and allowed "spirit energy" to flow through their fingers. As the planchette rolled, the pencil scrawled its way across the paper, sometimes tracing meaningless scribbles and sometimes what appeared to be written messages.

The medium, Crookes observed, shut her eyes, placed her fingers on a planchette, and waited for it to slide. As the chemist told the story, he was prepared to be entertained.

Could this invisible spirit, Crookes asked, see everything in the room, including things the psychic could not?

"Yes," wrote the planchette.

Crookes stepped backward. He had noticed the *Times* tossed upon a small occasional table.

"Can you see to read this newspaper?"

"Yes," was the reply of the planchette.

" 'Well,' said I, 'if you can see that, write the word which is now covered by my finger and I will believe you.' "

Standing with his back to the newspaper, he reached behind him, and pressed the tip of his right forefinger down on the newsprint.

The planchette hesitated under the woman's fingers.

"Slowly and with great difficulty the word 'however' was written. I turned round and saw that the word 'however' was covered by the tip of my finger."

Crookes wasn't the kind of man to be coy about his findings. In the winter of 1871 he published his planchette story—and his conclusion that it suggested some inexplicable power—in the *Quarterly Journal of Science,* a publication at which he served as editor. A few months later, Crookes published a more detailed and even more provocative report, the results of a series of tests he'd conducted on Daniel Dunglas Home.

The elegant Home still maintained his position as the professional medium most annoying to the scientific community. He'd risen above setbacks and controversies: Browning's vitriolic portrait of him as Mr. Sludge, a very public lawsuit by a patron who wanted her money back, accusations that he was merely a talented magician, a proposal by an anthropologist that Home was actually a werewolf "with the power of acting on the minds of sensitive spectators." The publicity always seemed to turn to his favor. His reputation flourished. In 1869, a trio of witnesses reported that Home had levitated off the floor of a second-floor room in a downtown London home, floated out from a window into the night, and drifted back in through the window of another room, his shadow flickering on the walls before his body slipped through the opening.

Home's numerous critics had several years earlier asked Michael Faraday to debunk the great medium, as Faraday had debunked talking tables. The irascible physicist had agreed, but demanded that Home first sign a statement saying that even if Faraday found that his results were "miracles or the work of spirits," the medium would admit that everything beyond those experiments was still "contemptible and worthless to mankind."

Home's supporters had shut down all further discussion with Faraday, and ever since, the famed medium had avoided the mainstream British science establishment. But at Wallace's urging, Home agreed to allow Crookes to set up some tests in the chemist's home laboratory and agreed that he would have no chance to inspect the laboratory in advance.

For one test, Crookes stretched a piece of parchment across a wooden hoop, attaching the hoop to a device that would etch any movement into smoked glass. He positioned Home near the experimental drum, not touching it but close. Crookes then took one of the medium's hands and held it a careful ten inches above the parchment. A friend of Crookes held Home's other hand at a similar level. The surface began to jiggle, pulling down, springing up. It looked as if an invisible hand was pushing at the surface of the parchment, lightly beating the drum, etching an uneven record into that glass plate.

In a similar test, Home was asked to stand by a mahogany board, set horizontally, but not to touch it. One end of the board rested on a pivot. The other was suspended from a spring balance. To guard against cheating, Crookes asked volunteers to hold the medium's hands and feet.

"Can you make it move?" Crookes then asked.

He'd attached a measuring instrument to the board, so that each movement would be again scratched onto a smoked glass plate. Home stood silent, Crookes reported, just stood there, hands and feet held tight, while the board rose and fell and the needle scratched a jagged line, peak after valley after peak, across the darkened glass.

As a result of those and other experiments, Crookes reached a conclusion shocking to his fellow scientists: that some kind of as yet unexplained "psychic force" existed, and that "of all persons endowed with a powerful development of this Psychic Force, Mr. Daniel Dunglas Home is the most remarkable."

In the next issue of London's *Quarterly Review,* a journal known for its dedication to the political status quo, came the response from the leaders of British science.

The article was unsigned, but its author claimed to know Crookes well as a scientist of "purely technical" ability but with a sad lack of real intellect. The writer then added, with finely honed innuendo: "We speak advisedly when we say that the Fellowship of the Royal Society was conferred upon him with considerable hesitation." Innuendo rather than truth; in reality Crookes, after he discovered thallium, had been elected as a fellow of the Royal Society by unanimous vote and, it had been noted at the time, without debate.

Like Wallace before him, Crookes was naive about how his colleagues might see his paranormal investigations. He'd expected demands for replication, perhaps competition from those who wanted to conduct their own studies of Home. He had anticipated criticism of the equipment he'd used, suggestions for better tests. He'd not expected to be slandered by anonymous report—or to see his friends slandered with him. Crookes had received help from an electrician named Cromwell Varley in designing and building the equipment used to test Home. The forty-two-year-old Varley worked as a consultant to the Atlantic Telegraph Company and had just finished working on the 1869 installation of the first trans-Atlantic cable, a feat that made telegraph communication possible between Europe and North America.

The anonymous article in the *Quarterly Review* (widely known to be authored by a physiologist and friend of Huxley's) also warned readers of "grave doubts" concerning Varley's ability, serious enough to prevent the electrician from being admitted to the Royal Society. At the time of publication, however, Varley had been a member of the society for three months.

Enraged, Crookes stormed down to the journal offices, demanded a retraction, and demanded that the author be publicly identified. The editor of the journal refused, explaining that it wouldn't be fair to the reputation of the writer, although as Crookes couldn't help but notice, the *Review* seemed unconcerned with his own reputation.

The message seemed clear enough. Investigating supernatural events was off limits to scientists, unless the findings proved fraud. Those who chose to ignore that rule—unspoken but strictly enforced—would find themselves off limits as well.

IT WAS A December evening in 1871. Myers and Sidgwick were shivering their way across the Cambridge campus. The air was as clear and chill as ice water. They walked under a sky netted with stars, those tiny silver points of light, so distant, so unreachable.

Myers felt that they glittered with mockery. He had moved no closer to understanding the secrets held so closely in the reaches of space. He had not resolved his doubts about life and death, achieved no means, it seemed, of finding answers. He had continued to delve into theological writings. Both he and Sidgwick had read Crookes's articles and the response to them. As far as Myers could see, traditional science and religion seemed interested only in proving themselves, not in resolving confusion, much less the questions of a lone British classics scholar. He felt completely on his own.

As he wrote years later, he kept looking back at those starry bits of light in the dark, looking up, looking out, and seeing not just the stars but a vast-ness that, in that moment, spoke of all that he wondered about the limits of physical reality. If there was an existence beyond the obvious, the mortal, shouldn't somebody be seriously, systematically trying to learn more about it? Shouldn't he be that somebody—he and Sidgwick, and perhaps a few others, like their mutual friend and fellow Cambridge scholar Edmund Gurney? The risks that such an effort would carry were apparent. One had only to observe the reaction to William Crookes's experiments to see how career-damaging an exploration into the unexplained might be.

Yet despite these misgivings, Myers turned to Sidgwick and asked whether the older man thought it was possible that by investigating "observable phenomena"—ghosts, spirits, whatever they might find—some valid knowledge could be uncovered.

Sidgwick, swathed in his dark coat and long scarf, was watching the stars, too. They seemed steady out there, timeless. He'd been thinking about what Crookes had done, thinking about the implications. As they stood wrapped in darkness, he made a decision. Yes, he told Myers, he did think a serious study of spirits might offer "some last grounds of hope" in reaching beyond the material earth and into unexplored worlds beyond.

Both men knew that whatever study was going to be done, the mainstream research community was neither willing nor prepared to take on this subject. So they would do it themselves.

Myers exhaled. He wasn't sure if it was the sound of relief or anxiety. He wasn't in the least sure what they would find. But he knew they were committed to looking. Sidgwick, once decided, was a man who always followed through. Myers hoped to measure up. "From that night onward," Myers wrote, "I resolved to pursue this quest, if it might be, at his side."

He was optimistic about their chances of success. But then, as he admitted to himself, he had always been the dreamer in the group.

3

LIGHTS AND SHADOWS

*F*UELED BY AN angry sense of purpose, William Crookes deliberately expanded his work in the supernatural realm. He methodically catalogued the phenomena he'd now observed: raps, levitations of people and objects, lights and luminous hands, and the very rare appearance of phantoms. According to Crookes's journal, he'd seen the latter only with D. D. Home, who could sometimes apparently persuade the shadows themselves to coalesce into human shapes.

Crookes emphasized that sittings with Home were rarely consistent. Over a period of hours, the medium alternated between producing startling effects and producing nothing at all. Crookes hypothesized that a greater power generated the good results, the "psychic force" that he had earlier proposed as the source of the medium's abilities. He thought Home possessed only erratic control over this force. Crookes's notion was that, as a talented psychic, Home either created or tapped into that unusual energy, which depleted itself, recharged, and was again exhausted during the sittings. As he made clear in his 1874 paper, Crookes valued the medium not as a performer but as a means of demonstrating that an occult force existed and might be measured.

In his more mainstream, scientific pursuits, Crookes was also fascinated by sources of energy invisible to the human eye. Even as he conducted his studies with D. D. Home, Crookes worked to refine the design of a piece of laboratory equipment used to observe a still mysterious kind of radiation. This device—which would come to be known as a "Crookes tube"—consisted of a sealed glass bottle, with almost all air removed, and with two electrodes set inside. When a high-voltage electric current passed between the electrodes, the tiny amount of air in the tube would begin to glow. If Crookes then removed all the air, creating a vacuum, a patch of fluorescence would appear on the wall of the tube. Because the light patch was aligned with the negatively charged electrode (called a cathode), the mysterious stream of luminous energy was named a cathode ray. Crookes continued to experiment with this strange radiation, eventually proving that the "rays" traveled in a straight line and could be blocked by objects in their path. Cathode rays themselves remained mysterious for more than a decade until the physicist J. J. Thomson—also using the controlled environment of a Crookes tube—deduced that the rays were streams of highly charged particles that Thomson called electrons.

At the time of his invention, and of his experiments with Home, neither Crookes nor his colleagues had any understanding of the structure of the atom, or of the energy embedded in subatomic particles like the electron. But the sense of natural sources of energy, of physical explanations waiting to be found, simmered in the science community. And it was that awareness of powers yet to be discovered that William Crookes found so compelling when he raised the possibility of a "psychic force" and urged his colleagues to abandon earlier misunderstandings and join with him in the search for it.

Crookes was such a smart scientist, and his reputation still so solid, that his psychical experiments proved difficult to dismiss, even among skeptics. Despite the attacks on his competence, and even his sanity, Crookes knew he was, at least, unnerving some members of the research establishment. Even Charles Darwin, after reading Crookes's reports, confessed himself "a much perplexed man. I cannot disbelieve Mr. Crookes' statement, nor can I believe his result." In genuine dismay, Darwin turned to his friend Huxley for reassurance. Huxley was quick to reply. He did not doubt that a talented conjurer—which he thought defined the mysterious Mr. Home—

could fool even a talented scientist. That, he could promise Darwin, was the entire story of the Crookes reports.

Darwin replied with real relief; he disliked wobbling away from the solid ground of scientific reality. He was now content in the knowledge, Darwin told Huxley, that "an enormous weight of evidence would be requisite to make one believe in anything beyond mere trickery."

In fact, Crookes also agreed that it would take an enormous weight of evidence to accumulate anything like belief. But his response was to advance rather than retreat. There was no one better qualified, he insisted, to dismantle the specious claims and to decipher the credible ones. He wanted researchers who believed, as he did, that "the supremacy of accuracy must be absolute." If Darwin and Huxley demanded hard evidence, then—in Crookes's view—unimpeachable researchers were required to cast the "worthless residuum of Spiritualism" away and reveal the promising remainder.

THE SAME YEAR that Crookes issued his call for help, John Tyndall, newly elected president of the British Association for the Advancement of Science, made it evident that no reputable researcher would follow that lead. Tyndall had been among the first to reject Alfred Russell Wallace's claims, and he continued to believe, passionately, that there was no room in the halls of science for even the possibility of spirit phenomena, ghosts, or gods. To think otherwise would have been to betray the new worldview, a worldview defined exclusively by scientists like Tyndall himself.

Tyndall was Faraday's successor at the Royal Institution; he had been elected to the association's presidency in 1874. He was an extraordinarily gifted scientist—and an equally formidable personality. Over the course of his career, Tyndall showed that gases could both absorb and transfer heat (a fact that would underlie theories of global climate change a century later); he demonstrated that ozone was a cluster of three oxygen molecules (others had mistakenly thought it was made of hydrogen); and he invented breathing devices for firemen and instruments to measure air pollution—notably the smoky fogs that too often enveloped London in choking darkness. His scientific prowess, combined with his new office, made Tyndall's a particularly persuasive voice as he evangelized for science and against any notion of a reality beyond the physical.

It was the age of scientific miracles, after all. Over the past decade alone Louis Pasteur had invented pasteurization, Alfred Nobel had invented dynamite, and Thomas Edison had invented the stock ticker. Tyndall's peers had in a few short years advanced civilization beyond anything, in his opinion, that religion had accomplished over many centuries. The new president of the British Association found it easy to dismiss biblical teachings as the explanations of yesterday. Any concept of an afterlife he jettisoned as unwanted, unwarranted dreaming—an illusion perpetrated on "the weak mind of man."

In his 1874 presidential address in Belfast, Tyndall urged spiritual leaders of all sorts—including the mainstream Christian clergy—to step aside: "The impregnable position of science may be described in a few words. We claim, and we shall wrest from theology, the entire domain of cosmological theory. All schemes and systems which thus infringe upon the domain of science must, in so far as they do this, submit to its control, and relinquish all thought of controlling it."

THE THIRD OF the founders of the psychical research movement, Edmund Gurney, was, at first, the most reluctant. Gurney possessed the outward appearance of a man lavishly blessed by life. He stood nearly six feet four inches tall, blue-eyed and fair-haired. He had an elegantly chiseled face, adorned by high cheekbones and a drooping handlebar of a mustache. He wore his top hats high and his tailored suits with style.

Born in Surrey in 1847, Gurney had inherited enough money to enable him to live in luxurious idleness—if he chose. He was as impatient with that prospect as he was with those who assumed his ambition reached no higher, among them the pretty woman who would become his wife.

He possessed "a mind as beautiful as his face," said the British novelist George Eliot, who belonged to Gurney's wide circle of friends. The rumor in London society was that the often-imperious Eliot had modeled the sweet-natured hero of her last novel, *Daniel Deronda,* after Edmund Gurney. She didn't deny it.

He could seem almost annoyingly perfect, until one got to know him better—and found the shadows under the gloss.

Gurney met Sidgwick and Myers while studying law and philosophy at Cambridge, a career he'd chosen after giving up his dream of becoming a concert pianist. They bonded over a shared love of philosophical arguments and of poetry, especially for that reigning poet, Alfred, Lord Tennyson. Hostesses of university parties spoke of them as an inevitable threesome— quiet Sidgwick, talkative Myers, and charming Gurney.

Myers was the only one of them who really enjoyed the circuit. Sidgwick was too shy to relax at parties, and Gurney, for all his social graces, too driven. "He could not bear to live without hard work," Myers recalled. There were reasons beyond ambition for Gurney to seek a working environment. The discipline of it seemed to steady him. All three men were intense by nature—another connection—but Gurney was unpredictably so. He had a mercurial temperament, rising to enthusiastic warmth, tumbling into chilly withdrawal, heating up again. Paralyzing bouts of depression interrupted his university career, necessitating a fifth year of study at Cambridge's Trinity College before he achieved his bachelor's degree. In a later century, his mood swings might have led to a diagnosis of manic-depressive personality or bipolar disorder. In the mid-nineteenth century, Edmund Gurney was simply a man given to extremes.

Despite his affection for his friends, Gurney was wary of their new direction; even at a distance, he could see how consuming it might become. So when Sidgwick first invited him to help investigate the spirit world, Gurney merely wished him luck. "Gurney will give us his warmest sympathies but no more," Sidgwick wrote to Myers in the spring of 1874.

Even Sidgwick's companions tended to underestimate his ability to gently and thoroughly erode resistance. Sidgwick politely acknowledged Gurney's rejection and continued sending information on the subject, which he knew his friend was too well mannered to ignore. Within weeks, Gurney's desk was stacked with a pile of well-selected, recently published arguments. They included Crookes's detailed accounting of his D. D. Home experiments and yet another plea from Alfred Russel Wallace for his fellow scientists to investigate "those grand mysterious phenomena of the mind, the investigation of which can alone conduct us to a knowledge of what we really are."

When Myers next visited him, in the seductive days of a late golden fall,

Gurney had changed his mind. Sometimes Myers worried, in retrospect, that he had bullied his gentler friend into joining this improbable crusade. But in that, he was mistaken. Gurney had awakened to the cause.

Two years earlier, Gurney had lost three younger sisters, drowned in a boating accident on the Nile River. The event had stirred a kind of inner rebellion. He was an educated man; he understood and even appreciated the arguments for a purely mechanical universe. Life lived as a cog in a cold, godless, indifferent machine, however, had come to seem to him unbearable.

As he would write to William James some years later, the "mystery of the Universe and the indefensibility of human suffering" were never far from Gurney's thoughts. He had no idea if scientists were wrong in their precise definitions of life's finite limits. But reading Wallace's arguments, poring over Crookes's deliberate experiments, Gurney realized that he wanted a chance to find out.

The three partners conducted their first serious investigation in early 1875, looking into claims that a pair of young mediums—teenage girls from Newcastle—could outdo D. D. Home in terms of materializations, summoning misty ghost children to appear from their cabinet.

As the investigators discovered, the Newcastle mediums and their regular audience of enthusiastic spiritualists clung to a certain routine. The mediums would retire to their cabinet, a curtained-off corner of a room, and allow their hands and feet to be tied while they reclined on couches. After the girls were settled, sitters would gather outside the cabinet. The mediums demanded a dark séance, with almost no illumination. The investigators sat in a dusky gloom with the gaslights turned down to a pale, blue glow. Myers complained that the gloom was so dense that he could barely see the curtains a few feet away.

And the Newcastle spiritualists insisted on singing. They began each séance with hymns, continuing for hours, whether or not a spirit arose out of the dusky murk of the room. Sidgwick finally begged them to allow him to quote poetry instead. It took two hours of Swinburne, beginning with the poet's "Atalanta in Calydon"—"Love that endures for a breath, Night, the shadow of life, And Life, the shadow of death"—before a pair of gauze-draped children drifted out of the cabinet. Sidgwick was capable of more—he had once recited poetry through an entire crossing of the English Channel—but when he next returned the spiritualists intercepted him, breaking into song as he entered the room, loud enough to drown out any recitation.

Meanwhile, the pretty little girl ghosts tended to gravitate admiringly toward the handsome Gurney. One evening, according to Myers's notes, a veiled spirit kissed Gurney two or three times and then, apparently feeling this inadequate, paused to "materialize her lips," peeled back the veil, and delivered another kiss with uncovered lips. Although Gurney reported the kiss as a solid smack, the spiritualists contradicted him, insisting that the affectionate ghost was no more than empty air under her veil. Obviously, there was no way to conduct an objective study while surrounded by enthusiasts. Myers, Sidgwick, and Gurney decided to evaluate the mediums on neutral territory and moved the whole investigation to London.

They met at the home of a new member of their group, Arthur Balfour, a wealthy aristocrat—and future British prime minister—who lived in the elegant neighborhood of Carleton Gardens. Balfour set aside a small room for the spirit cabinet and kept the door locked. When in the room, the mediums were confined by a harness of leather straps and combination locks to the wrought iron surround of a fireplace. They were allowed a drawn curtain to provide the closed "energy" of a cabinet.

The investigators neither sang nor quoted poetry. It had occurred to Sidgwick that those long, tedious hours of singing allowed the girls, especially if they had a confederate, to slip, or break and replace, any ties. Meanwhile, the noisy harmonies would have masked such deception.

So in London, the investigators just waited. And waited some more. It was hard to say whether the mediums or their watchers were more stubborn about it. But by the twelfth night, only Sidgwick and Balfour's two sisters, Eleanor and Evelyn, were still wearily observing the limp curtains that screened the mediums. At long last, a small white form appeared. Squint as he might, Sidgwick could not convince himself that he was looking at a spirit. "We all thought the movements of the small figure just like those of a girl on her knees," he wrote Myers. Even worse, Eleanor Balfour had asked to search the mediums. They refused to allow her near them. The whole Newcastle production was clearly just that, a production.

Curiously, Sidgwick was not depressed. As he once told Myers, he approached spirit phenomena much as he did religion. "I believe there is something in it; don't know what." He was determined to see past the "obvious humbug" to what might actually exist. This time, if nothing else, the investigators had proved themselves reassuringly capable of ferreting out fraud. And he had become acquainted with Nora Balfour.

"FRIENDSHIP BETWEEN the sexes is, you know, after all a devilish diffi-cult thing," Sidgwick wrote to a friend that summer. "How are you to pre-vent mistakes on one side or the other?"

He feared nothing more, at that moment, than the humiliating mistake of unrequited love. Sidgwick's attraction to Nora Balfour had continued to grow. Every time he met her, he liked her more. But how did one approach a woman so perfectly contained, so elegantly composed, so concealing in her nature?

Nora Balfour turned thirty in 1875. She was a slight woman, straight in posture, crisp in manner of dress. Her hair was a pale ash brown, pulled severely back from a thin, fine-boned face. Her eyes were ice-water silver, and her voice as cool as a winter's day.

She rarely laughed, never joked, loathed small talk. But she was amaz-ingly, fearlessly, openly intelligent. One of her favorite pastimes was doing mathematical calculations with her sister's husband, already a noted physi-cist, John Strutt, Lord Rayleigh.

Rayleigh had married Evelyn Balfour in 1871, the same year he'd published his famous answer to a famous question: Why is the sky blue? His answer—because of the way light bounces off small atmospheric molecules—was known as Rayleigh scattering.

Since then the baron, working from a laboratory at Cambridge, had been running calculations on the concentration of atmospheric gases, try-ing to bring some order to all that shifting color and energy swirling around the planet. He'd recruited the talented Nora to help him. On a family vacation to Egypt, while the others went shopping and touring, Nora and her brother-in-law spent every morning happily doing math together, with the shades drawn against the glowing Egyptian sunlight.

Like Rayleigh, Henry Sidgwick admired smart women. But Nora was so outwardly self-contained that Sidgwick had no idea of her affections. From her countenance, she might be politely indifferent, mildly friendly, or—he could only dream—simmering with hidden passion.

"It is not as if the human heart was only capable of the one or the other definite emotion," he complained. One could make analogy to colors—call affection blue, he said, and call love red. If there were only such distinct colors to emotions —"blue or red: then it would be comparatively easy to distinguish what was proffered. But, on the contrary, there are all sorts of purples which run into one another. . . ."

Fortunately, his spirit studies gave him a reason for frequent visits. The Balfours, as a family, approached the investigations much as Sidgwick did—with a belief in the importance of the work and an acknowledgment that it had barely begun. Rayleigh himself best summarized the consensus: he found the few good experiments, notably William Crookes's work with D. D. Home, compelling enough that he was "prepared to be converted." But, Rayleigh added, he had not yet seen the evidence that would effect such a conversion. He needed more.

As the summer of 1875 deepened into autumn, over discussions of ghosts and morals, work and life, Sidgwick discovered that he could detect the color of Nora Balfour's feelings and that they were, after all, a strong and joyful red. In a private letter, she assured him of her love. In December, she accepted his proposal of marriage. "Everything is always better than it is expected to be," Sidgwick wrote to Myers. As in a romantic ballad, the croakiest bird sounded of golden song; the few remaining, weedy flowers spilled perfume into the air, ever since "my happiness began."

Rayleigh had sent him a note of congratulations: "You lucky dog!" it read.

The course of spiritual research, though, was running a little less happily. For all the insistence of Wallace and Crookes that only scientists could do the work dispassionately, neither man seemed unusually objective to the Sidgwick group. They seemed, rather, to be as human as anyone else.

Wallace had once told Darwin that he never could resist "an uphill fight in an unpopular cause." He now appeared determined to push the limits of unpopularity, defending practices that even the poetry-loving philosopher Sidgwick thought dubious at best. Among these practices was the new fad of "apports," objects that mediums claimed to be able to materialize with the help of friendly spirits. An apport might consist of an apple, a pile of pink shrimp, a fluttering dove, or, in Wallace's case, a veritable shower of fresh flowers.

In one sitting, held in midwinter in Wallace's home, a wild bouquet suddenly appeared on a bare table: anemones, tulips, chrysanthemums, Chinese primroses, and ferns. "All were absolutely fresh as if just gathered from a conservatory. They were covered with a fine cold dew. Not a petal was crumpled or broken, not the most delicate point or pinnule of the ferns was out of place." And their scent was so strong in the warm, gaslit room, Wallace added, that it would have been impossible to have concealed them there in advance. He did not attempt to explain why spirits would want to strew a room with blossoms, except perhaps as a demonstration of their powers. But he challenged his fellow researchers to prove such events fraudulent.

No one doubted Wallace's sincerity, but his colleagues began to perceive in him an urge to provoke. There was no doubt that provocation rang like a bell in his recent declaration that the facts of spiritual phenomena "are proved, quite as well as any facts are proved in other sciences."

The problem with Crookes took a different shape—a feminine one. Sidgwick and his friends were coming to suspect that the chemist was measurably more objective about male mediums than about their female counterparts.

Fred Myers was the first to perceive the risk. He'd gone to London to investigate a very pretty new medium, a blond American named Anna Eva Fay, who performed in Davenport Brothers style with a showy cabinet, ghost hands, and mysterious music. Crookes had immediately warned Myers off, saying that he had priority claim on such investigations.

"The lion will not be robbed of his cub," Myers wrote to Sidgwick, adding, "nor the cub her lion." Fay visibly appreciated Crookes's protection, as well she might. The resulting experiments seemed unusually accommodating. The medium asked to work in a darkened room. Crookes, in a reversal of his earlier insistence on light with D. D. Home, agreed to that request. He also quickly announced the tests successful.

There was no doubt, Myers wrote to Sidgwick, that the newcomer was beautiful and charming. He himself found her delightful. He just wasn't sure she was trustworthy. Myers persuaded Fay to visit Balfour's home, for just a few little experiments. Actually, he had one in mind—simple immobilization. They tied the medium's hands to widely separated metal rings, fixed to a board, and then waited for the "spirits" to help her out. The observers waited until it became obvious that nothing was going to happen. When Myers went home that night, he pulled out his journal and drew a fat black line through Fay's name. Equally irritated, Sidgwick proposed that they begin avoiding professional spiritualists.

It appeared, he said, that paid mediums brought nothing to the enterprise but "persistent and singular frustration." They could only hope that a professional scientist, like Mr. Crookes, would be as wary. But the Sidgwick group was reaching a new awareness—that in their smug sense of superior intelligence and capabilities, trained scientists did not always see what was obvious to others.

EVEN THE FAMED spiritualist medium D. D. Home now warned that the profession was rife with fraud, to the point that he felt a need to provide some exposure. In particular, Home decried the increasing use of "dark séances," conducted in a gloom so dense that no one could follow a medium's actions, a darkness justified by the claim that spirits preferred the shadows. "Light should be the demand of every spiritualist," wrote Home in a book he completed in 1876. "By no other [tests] are scientific inquirers to be convinced. Where there is darkness there is the possibility of imposture—and the certainty of suspicion."

Home's tuberculosis symptoms were creeping back, sending him to the south of France in hopes that a softer climate might prove healing. His days of calling up spirits and floating off the floor were over. Ever a man with a mission, Home had decided on his own last act—to expose and help remove fraud from his profession.

In *Lights and Shadows of Spiritualism,* written with the help of his longtime lawyer, Home took a fierce stand against not only dark séances but the wide range of conjuring tricks used by his fellow mediums. He went on to explain some of them in detail, including the way to fake "full-form materialization," the type favored by the unfortunate Newcastle mediums. According to Home, a medium only need conceal fine muslin fabric—the filmy stuff produced in France was a favorite—in her underwear. She could count on the fact that no gentleman would invade a woman's privacy by searching there. After retiring into her cabinet, the medium could wiggle out of any knots, strip, lay her dark clothes on pillows arranged on her resting couch, and flit out into the room draped in her floating muslins.

For the first time, Crookes was genuinely unhappy with D. D. Home. His wife wrote to the ailing medium, saying she feared his revelations could do further harm to her husband's reputation: "Of course, it is looked upon as a complete exposure of the whole subject by the biggest imposter of them all," she wrote, adding that Home's book was "the best weapon" their enemies could acquire. She and her husband had reason to feel defensive by then; Crookes had become involved with another female medium, moved into another "lion and cub" relationship, this time with one of London's most artful psychic performers. His reports on Florence Cook were so smitten that the near universal verdict was that she had become his mistress.

Barely in her twenties, Cook was strikingly pretty, with big dark eyes, thick, curly hair, and a curvy body that she liked to emphasize with closely fitted black dresses. Of course, those black garments and matching black boots made her hard to see in the dark, which was where she liked to hold her séances, comfortably ensconced behind the silken curtains of her cabinet. What made Cook noteworthy was that while she was tied up in her cabinet, her "spirit guide" would wander into the room.

Almost all working mediums claimed to need a guide, a spirit to help them navigate the mysterious realms of the other world. Trance mediums called their guides "controls," giving their trances a rather eerie suggestion of possession. When a medium was in a trance, the control would speak through her or him, summon other spirits, and impart advice to those gathered around. Cook claimed to be able to coax her guide right into the room, fully materialized as a graceful young girl named Katie King, a creature of flowing white robes and charming manners.

Katie King could flirt, eat (cakes), drink (wine), and tuck her plump little hands into William Crookes's arm. The scientist confessed himself enchanted. He'd taken pictures of the lovely spirit, but "Photography is as inadequate to depict the perfect beauty of Katie's face, as words are powerless to depict the charm of her manner."

But Florrie Cook didn't move in the elevated circles where D. D. Home held his exclusive séances. Until Crookes gave her special status, Cook had been just another pretty street medium. She moved in a circle where violent competition was the rule, and her resentful competitors were quick to bring her back down to their level. Stories spread that Cook was having an affair with Crookes, that he was more lover than scientific observer. One mother-daughter pair of jealous mediums forged letters to confirm the illicit relationship.

Crookes denied the charges vigorously. But he was increasingly aware that his explanations were dismissed and his denials brushed away. It was to D. D. Home that he then confided his sense of betrayal, writing a letter that overflowed with hurt pride and genuine hurt as well. As Crookes told Home, "were it not for the regard we bear to *you*, I would cut the spiritual connection, and never read, speak or think of the subject again." He had decided not to hunt for another Home, at least not any time soon. Instead, Crookes determined that it was time for him to return to the safer—and definitely saner—world of mainstream science.

EVEN AS CROOKES WITHDREW, though, another conventional scientist, the first physicist to really embrace the idea of occult studies, made an unexpected foray into supernatural research.

William Fletcher Barrett had been educated under John Tyndall, that outspoken opponent of all things spiritual. Barrett had worked in Tyndall's laboratory for four years before taking a position as professor of physics at Dublin's Royal College of Science.

A big, tousle-haired bear of a man with a dramatic flair for telling a tale, Barrett may have lacked his mentor's dazzling genius, but he was a thorough and conscientious researcher who had assigned himself the task of sifting through the electromagnetic properties of iron alloys. Barrett had little interest in the supernatural, although he had earlier—in 1871—corresponded with William Crookes, who was then seeking a theory to explain his results with D. D. Home. "If you can help me to form anything like a physical theory, I should be delighted," Crookes wrote. Barrett, not particularly drawn to spiritualism, had casually written back to propose that hallucination might be a good answer.

A few years later, though, Barrett would change his mind about that hasty reply. He enjoyed a habit of spending part of each summer vacationing with friends in the countryside south of Dublin. His friends were fascinated by hypnotism— also called mesmerism, due to the early influence of Anton Mesmer on the field—and they occasionally conducted experiments with local villagers. These exercises amused Barrett; he held to the earlier idea that these so-called trances were some form of hallucination. It was only in 1875 that the tests, carried out with a group of village girls, caught his attention.

When hypnotized, one of the girls seemed sometimes to make an uncanny connection to another person's thoughts. Barrett, suspecting trickery, some kind of signaling, decided to lend a scientist's hand. He blindfolded the girl and then asked his friend to perform a selected series of interactions with the hypnotized girl.

As Barrett later wrote, "If [the friend] placed his hand over the lighted lamp, the girl instantly withdrew hers, as if in pain. If he tasted salt or sugar, corresponding expressions of dislike and approval were indicated by the girl." Occasionally, the girl seemed also to inexplicably know some of what Barrett was pondering as he ran the experiments. Not precisely, but close enough, a "more or less distorted reflection of my own thought."

When Barrett returned to his university, he conducted some follow-up experiments and drafted a paper on his research. He gave it the studiedly neutral title "On Some Phenomena Associated with Abnormal Conditions of Mind." He then submitted his paper for presentation at the 1876 meeting of the British Association of Science, the body that, headed by Tyndall, had urged all spiritualists, as well as traditional religious leaders, to step aside for the edicts of science. The association reviewers weren't fooled by the title; they knew crackpot science when they read it. Barrett's paper was rejected first by the biology division and then by the anthropology division and was scheduled for the trash pile when the chairman of the anthropology subsection, Alfred Russel Wallace, rescued it and used his position as chair to force it into the program.

In his paper, Barrett wrote with the caution of a man inching his way across thin ice. He tentatively proposed that at some unconscious level, it might be possible for a person to tap into another's thoughts or feelings. He speculated that this might occur in a "community of sensation," and suggested that further investigation might be warranted.

Barrett stood firm only in his conclusion, which sounded a warning note. If he and his researchers rejected phenomena simply because they seemed strange and so far inexplicable, scientists might end up "laying ourselves open to that same spirit of bigotry that persecuted Galileo."

Not unexpectedly, Wallace joined him in urging further investigation.

More surprisingly, at least to the members of the association, the decidedly science-centered Lord Rayleigh also urged further investigation. "My opportunities have not been so good as those enjoyed by Professor Barrett," Rayleigh explained. "But I have seen enough to convince me that those are wrong who wish to prevent investigation." Rayleigh might be unconvinced as yet himself, but he wasn't afraid to chastise his peers for taking a political rather than a scientific stance.

Tyndall was flat-out furious. It was outrageous that one of his students had gone so wrong and that the association, under his command, appeared to endorse this preposterous minority view. He wanted it made clear that no real researcher believed that there was anything, anything at all, in thought transference or "mentalism," as self-anointed psychics loved to call it. The scientific association's response came like a thrown missile: an investigation was out of the question. Further, Barrett's would be the one conference paper that would *not* be published in the meeting proceedings.

Unfortunately, from that perspective, Barrett's study could not be kept completely secret. The initial report received plenty of newspaper coverage and excited interest from readers, as well as rare approval from Sidgwick and even an offer of money from Fred Myers. "Remember that if in any way your experiments can be helped by money, I shall be positively glad to know of what I consider so good a way of spending some," Myers wrote.

Still, Barrett was now also angry. Although he had known that his report would stir controversy, Barrett had not expected to be treated so shabbily. He was Tyndall trained; he was thirty-two years old; he had spent more than a decade following the rules of science to the letter. He *knew* his study to be worthy. His experiments were careful, his arguments rational. Barrett was irate enough not to step meekly back into line. He decided to continue his investigations, with or without the support of mainstream science. And he chose not to tiptoe around the results this time. He chose to make his next study as public as possible.

Barrett fired his first salvo by writing a letter to the *Times* of London in September 1876, announcing that he was interested in mind-reading and therefore seeking letters from people who had witnessed a good illustration of "the willing game."

He described the response as "a flood of replies from all over England." His university office was buried in banks of white envelopes. Letters formed drifts across his home office like the remnants of a winter storm. The postman literally staggered to his door, dumping resentment and pounds of mail in equal measure.

Barrett's fellow scientists might not be interested in his new research project. But it seemed that just about everyone else was—or that a surprising number of British residents liked to play the willing game.

The willing game was a dinner party entertainment with a mind-reading twist—and a clever title. The rules were simple: after guests had gathered, one person would be chosen to leave the parlor. Once he or she had left, a task would be decided on: perhaps picking an iris out of a bouquet, maybe selecting a certain book from a shelf, kissing a wife, straightening a husband's tie, finding a hidden object.

Ideally, the chosen person would be "willed" into the action chosen, following directions sent mentally by others. Often, he or she would be blindfolded before reentering the room, guided by another player. Sometimes the guide would place a hand lightly against the seeker's shoulder or forehead, using touch as an additional aid in "willing" a job to be done.

To scornful physiologists, this was simply another version of table turning, easily explained by applying Faraday's golden rule of unrealized motion or subtle cuing. It was all "unconscious muscular action on one side, unconscious muscular discernment on the other." The guide was probably giving inadvertent signals, they said, tightening fingers in response to one move, pushing a little in response to another. The blindfolded follower was undoubtedly doing just that—following.

The willing game was an invitation to cheat, scientists said. It only required two confederates, agreed to work together in fooling the rest of the room. It was no wonder that so many wives were successfully kissed, irises plucked, books correctly selected off shelves.

When would the gullible public learn that it was participating in another variation on planchettes, tilting tables, and spirit cabinets? At least the willing game, unlike the tricks of professional, paid mediums, was played in private, with no money at stake.

In the case of slate writing—another new technique for supposedly communicating with the dead—one eminent British zoologist, Ray Lankester, became so outraged by the gall, the effrontery of it, that he decided to expose its leading practitioner—just as publicly as Barrett had with his mind-reading investigation.

The medium in question was one more American, styling himself as the new D. D. Home. As such, Henry Slade cultivated an ethereal look. A London newspaper described him as possessing a "dreamy mystical face, regular features, eyes luminous with expression, a rather sad smile and a certain melancholy grace of manner." Like Home, he could also make tables levitate and ghostly hands float through the air.

But his claim to fame was slate writing, described as a talent for inducing "spirits" to write directly onto slates, without any need for a planchette-like apparatus. Slade used small rectangular slates, or chalkboards, much like the kind that children use for lessons in schools. At its most basic, slate writing involved holding a blank slate and chalk pencil under a table. The medium and the visitor would wait until they could hear the scratchy sounds of writing, the pencil apparently propelled by a mysterious power. When the slate was brought up for examination, messages would be scrawled across the formerly pristine surface.

Lankester wasted no time putting this nonsense to bed. He met Slade, sat down at the appointed table, watched only for the writing materials to be placed beneath it, and then yanked the slate back out, not bothering to wait for any sounds of scribbling. As Lankester wrote to the *Times,* the slate in his hand was already fully inscribed with a written message. In other words, he explained to the newspaper's readers, the medium had switched the first blank slate for one already containing a "spirit" message.

Lankester wasn't content with merely embarrassing the visiting performer; he insisted that Slade be prosecuted as a con artist. In the resulting October 1876 trial, Lankester also insisted that science dominate the proceedings, that the judge make a decision based only on "inferences to be drawn from the known course of nature."

Despite a few eloquent defenders—including Alfred Russel Wallace—Slade was convicted and sentenced to hard labor. To Lankester's disgust, however, the sentence was overturned on a technicality, and Slade fled to the Continent. Nevertheless, the scientist thought he'd achieved his goal to make it clear that researchers would no longer tolerate this endless series of murky and mystical claims. He hoped that the subject would be dropped, preferably forever, and that scientific discussions would no longer be "degraded by talk of spiritualism."

IN THE YEAR 1876, the year of Slade's trial, the year that introduced the player piano and the National Baseball League, the year the telephone was invented in America and the internal combustion engine in Germany, the year William James was named an assistant professor of physiology at Harvard, the year of Henry and Nora Sidgwick's marriage, Fred Myers's attention, heart and soul, belonged to a woman he could not have.

For Myers the winds of change blew dark, blew stormy, for 1876 was also the year that the woman whom Myers loved committed suicide.

He was riven by the belief that he should have known, should have prevented it somehow. The elusive Annie Marshall, with her unforgettable "sea-sapphire" eyes, had been desperately unhappy for years. Her life with husband Walter Marshall, Myers's cousin, had descended into despair along the spiral of his mental illness. The extended family had tried to support the couple. Myers himself had gone to Switzerland, in 1875, to find some new doctors for Walter, to comfort his wife, and—to their mutual dismay—to fall in love.

Myers filled his diary with poetry about her beauty, her sadness and the way his *"soul sprang to meet her . . ."*

After Walter's behavior became so violent that he was institutionalized, Myers's mother offered Annie a home, and Myers wrote her brisk, friendly letters. His instincts told him that the last thing Annie needed, at the moment, was a suitor. Resolved to hide his true feelings, he confided to his diary that he preferred "irredeemable woe to the slightest shadow of wrong."

But with Walter locked away, Annie fell apart. She seemed to blame herself, Myers's mother wrote him. The morning after she sent that worried letter, fretting over her houseguest's sudden silent withdrawal, Mrs. Myers could not find Annie Marshall in the house. She ran outside to look, eventually reaching a nearby lake, and to her horror found Annie's lacy shawl crumpled by the water's edge. Alongside the shawl was a bloody pair of scissors.

The body was found floating in the deep middle of the lake. When they pulled Annie out of the water, they realized that she had tried to slash her throat first, but the wound was too shallow to kill. In the despairing dark of early morning, she had then walked, bleeding, into the shallows, and on, until the lake closed over her head.

The news, Myers wrote, "made dim with woe the reeling world." He was wrenched with feelings of guilt and inadequacy. He began to believe that he must talk with Annie again—for both their sakes.

The following year, Sidgwick insisted that Myers go to Paris to explore the medium scene there. Frankly, he mostly wanted to put his friend back in motion; Myers had remained miserable, uncharacteristically antisocial and lethargic. Certainly, Myers complained about the summer 1877 trip to France. Life seemed discouraging to him after Annie Marshall's death. And, along with his friends, he was discouraged by their psychical results— or lack of results—thus far. But on his duty trip to France, for the first time, Myers caught the dark scent of the occult in the air.

He hadn't expected it, and it came without flourish. In one visit to a tidy home where a pleasant middle-aged French woman specialized in table talking, the knocks in the wood had suddenly spelled out a name— "Annie."

The startled Myers had demanded that the medium tell him what relationship this Annie had to him.

"Cousin," came the reply.

It was, Myers wrote to Sidgwick, perhaps a case of thought reading, the kind of unexpected insight that William Barrett had detailed in his experiments.

A few days later Myers met with a fat old medium in a filthy Paris bystreet near the buttressed walls of Notre Dame Cathedral. She wore a too-tight dress with its lace-edged skirts flapping around her ankles. She sat on a small wooden stool and leaned against a stone wall, holding a tray of wooden alphabet tiles.

The medium claimed that when sitters touched the right letters on the tray, the spirits rapped in response. As he confessed, Myers had to fight to keep a straight face. He decided just to have some fun with it.

"I despaired of getting a high spirit thro' this low woman & amused myself by laying traps—i.e. *thinking* of the right words but pausing & trembling at the wrong letters."

To his surprise, though, he found the letters that the medium picked out seemed to follow his thoughts rather than his actions. Slowly they stumbled through an attempt at a name: ANNE ELISE MARSEH PAR MARESHELL.

Standing in the grubby alley, Myers thought he felt the oddest little chill down his back, unnerving as it was invigorating.

Even so, even as Myers welcomed that shiver of hope, he was aware it might be time to let Annie go in favor of living companionship. He was surrounded by domesticity. Not only was Sidgwick happily married, but in the spring of 1877, Gurney had also given up bachelorhood.

There were hints that Gurney's pretty wife, Kate Sibley, had married for money. But the couple seemed happy enough together. Kate's quick chatter and light laugh made a bright counter to Gurney's quiet nature and occasional dark moods. She'd been keeping him unusually busy too, redecorating their home in London's Montpelier Square, coaxing him out to the theater and to dinner parties.

Myers found himself more alone than usual, less satisfied with pursuing an affair with a ghost. And there was one woman he did find charming, a dark-haired society beauty named Eveleen Tennant. She was looking for an older man to take care of her; she loved his enthusiasm, his intelligence, and his open-minded approach to the world. "With you," Evie told Myers, "I can be my wild, natural self." He was delighted by her beauty and passion. In his diary, Myers tended to sketch rabbits by her name. Six months after they met, in 1880, they were married in a formal wedding at Westminster Abbey.

Evie Myers had no reason to doubt herself. A talented portrait photographer, she was known for her images of the well connected and powerful. The National Gallery would eventually acquire much of her portfolio, including a rather dreamy portrait of Nora Sidgwick's brother, Arthur, during his political rise to prime minister and another of Robert Browning, glaring at the camera with the intensity of a trapped hawk. Yet for all her accomplishments, the artist clung to her new husband as if she were his needy child.

"You can live without me," she wrote in one overwrought letter, "but I cannot live without you." She sensed in Myers an emotional distance that troubled her. Did he compare her, she wondered, to women in his life who were more intellectual and independent? "Fred my king and my love," she continued, "I will try & be worthy of you but *do not* want me to be like your dear mother or Mrs. Sidgwick. I can't."

He reassured her, but he didn't tell her the truth. Of course he didn't want her to be like his mother or Nora Sidgwick. He wanted her to be like Annie Marshall. Long after they were married, after their three children were born, Myers would still look at his beautiful wife and want her to *be* Annie Marshall, apparently forgetting that it was so much easier for the dead to keep their perfection.

In 1881, after five years of conducting tests and compiling results in his spare time, William Barrett published his second report on the science of thought transference.

His results, summarized in the esteemed journal *Nature,* detailed test after test, done with a carefully chosen set of people, identified through their willing-game success or other evidence of mind-reading abilities.

Among them were three young girls, daughters of a Presbyterian minister, who seemed startlingly responsive to Barrett's mental instructions. Barrett assured readers that he took constant measures to prevent cheating in the experiments. For one series, he settled the girls in the family parlor and then closed himself up in their father's study. Once seated at the heavy oak desk, Barrett began slowly listing household objects on a sheet of paper.

The only instructions he gave the girls were to bring him the object that entered their minds. Only after they'd brought one object to the study—and he'd either accepted it or told the bringer to try again—would another command be put to paper. "Having fastened the doors. I wrote down the following articles, one by one, with the results stated: *hairbrush,* correctly brought; *wine glass,* correctly brought; *orange,* correctly brought; *toasting-fork,* wrong on first attempt, right on second; *apple,* correctly brought; *smoothing-iron,* correctly brought; *tumbler,* correctly brought; *cup,* correctly brought; *saucer,* failure."

Lucky guesses were certainly possible, but, Barrett thought, not to this degree of consistency. Further, there could be no accusations of secret nudge-and-wink techniques, as with the willing game, "for there was no contact, and in some trials (as in the foregoing) the percipient was out of sight and hearing." It was possible that the girls had cheated, but it was difficult to see exactly how. Barrett's requests had sometimes been impromptu, and when they weren't, he hadn't discussed them.

What he saw was not a solution to the mystery he'd raised earlier but rather more mystery, "a large residuum of facts wholly unaccounted for." There was a notable difference in this paper's conclusion, though. Barrett asked no help from the traditional scientific community; he no longer expected it. The ability to conduct such an investigation, he wrote, seemed to lie "outside the scope of any existing scientific society."

THE BRITISH SOCIETY for Psychical Research formally convened for the first time on February 20, 1882, representing a branch of science so new that the organizers had felt compelled to invent a name for it. They'd argued over the proper term, settling only with some dissatisfaction upon "psychical research."

"We could find no other convenient term," explained Myers, "under which to embrace a group of subjects that lie on or outside the boundaries of recognized science."

Henry Sidgwick, the first president, began on a chiding note, declaring that he and his colleagues had been left no choice but to invent a science, to create a support system for those who wanted to do the work. The SPR came into existence by necessity, Sidgwick said, founded because there were questions—of immortality and of humanity—that demanded investigation. And it was founded, he said, because conventional science had tried to block even the most modest of inquiries along those lines. The normally soft-spoken Sidgwick called that interference "a scandal to the enlightened age in which we live."

But he didn't waste undue time on the past; they all agreed that a formidable job lay ahead. Tasks were assigned, with Frederic Myers and Edmund Gurney responsible for investigating apparitions and William Barrett heading the thought-transference committee. Nora Sidgwick was put in charge of investigating ghosts, though privately confessing to her husband that she didn't believe in them.

William Crookes attended that first meeting, as did Alfred Russel Wallace, although the latter had expressed some concern over Sidgwick's too-skeptical approach to the field. In addition to Nora, the Balfour family was amply represented by Gerald and Arthur Balfour, their sister Evelyn, and her husband, Lord Rayleigh. Aside from the familiar supporters, the society also attracted an impressive array of new faces. The SPR's membership list grew rapidly to more than two hundred. It included painters, clergymen, politicians, spiritualists, and a cast of writers that ranged from Alfred, Lord Tennyson (at the time Britain's poet laureate) to Leslie Stephen, editor of the voluminous and influential *Dictionary of National Biography* (and father of a baby girl, born less than a month earlier, who would become the novelist Virginia Woolf). The society also attracted the essayist and social critic John Ruskin and the Reverend Charles L. Dodgson, who wrote his *Alice in Wonderland* and *Through the Looking Glass* fantasies under the pseudonym Lewis Carroll.

The American writer Samuel Clemens joined as well. Clemens had gained international acclaim for his 1876 novel of unruly boyhood, *The Adventures of Tom Sawyer,* written under the pen name Mark Twain. The novelist had a specific purpose in joining, entirely removed from his chosen career. He wanted an explanation for a dream that had haunted him for twenty-four years:

> In 1858 the Clemens brothers, Sam and Henry, were training to be riverboat captains, working the Mississippi River together on a big, steam-powered paddle-wheeler called the *Pennsylvania.* On an early June evening, the boat docked in Saint Louis, and the brothers went ashore to visit their sister. After dinner, Henry went back to the *Pennsylvania.* Sam stayed the night at his sister's house.
>
> Just as Sam Clemens started to slide into sleep, an image formed, a horrifyingly detailed dream in which he saw his younger brother's body tucked into a casket. The coffin lay balanced across two chairs. Flowers sprayed across Henry's unmoving chest, a cascade of white roses with a single red bloom in their center.
>
> Samuel Clemens sat up in bed, gasping, his heart pounding. He stumbled downstairs, half awake, the dream still so real that he was braced against the sight of his brother's body in the parlor. He'd been almost shocked to find the parlor quiet and dark, its chairs empty of dead men, its air unscented by roses.

Just a dream, he told himself, just a dream.

When Sam returned to the *Pennsylvania* that morning, his brother was waiting—whole, healthy, a little sleepy in the morning light. But they were separated again; the captain transferred Sam over to help on a companion boat, one that trailed behind by a day. Three days later, the *Pennsylvania*'s boiler exploded, just as the boat cruised south of Memphis. One hundred and fifty people were killed or injured. As soon as the news reached him, Sam Clemens left his boat, hired a fast horse, and rushed to Memphis, where survivors were filling the local hospital.

Henry Clemens died that night with his brother sitting beside him. In the morning, Sam walked numbly down to a room where the bodies of the dead were awaiting burial. Henry lay in a metal casket, balanced across two chairs.

As Sam Clemens stood, blinking against the memory of his dream, a volunteer nurse stepped up to the coffin and gently laid across it a bouquet of white roses with a single red bloom in their midst.

A DREAM, a coincidence, a premonition, a horrifying glimpse into the future? The older brother had wondered for years. Now, Clemens hoped, the time had come, the commitment was there—and he would finally get an answer.

4

METAPHYSICS AND
METATROUSERS

THE AIR WAS CLOUDED black that fall of 1882, when William James came to London, barely six months after the Society for Psychical Research first convened. A miasma of fog and the smoke of tens of thousands of chimneys shrouded the city. It was, James said, like living in "the inside of a coal mine—and a coal mine in the process of combustion at that."

He arrived on a Harvard-approved sabbatical, taking a break from teaching and family to focus on his ambitious plan for a new textbook in psychology. James had signed the contract to write the book in 1878, the same year that he'd married a dark-eyed, soft-voiced Boston schoolteacher, Alice Gibbens Howe, and begun family life in a small Cambridge home.

In that autumn of 1882, the couple already had two sons—three-year-old Harry, whom his father liked to refer to as "my domestic catastrophe," and three-month-old Billy. But for all that—or perhaps because of all that—William was frustrated with his work. Psychology was a slippery field to catch hold of, a science just defining itself. James liked to joke that "the first lecture in psychology that I ever heard was the first I ever gave." He'd started teaching psychology at Harvard in 1875, inventing the coursework as he went along.

The idea of a science of the mind had been gaining power since the turn of the nineteenth century, built on early studies of brain anatomy, dissections of the spinal cord and nerves, and awareness that brain damage could directly affect behavior. In 1848, a French neurologist had offered 500 francs to anyone who could show him a brain from an individual who suffered from speech disturbance and did *not* have damage to the left frontal lobe. By the next decade, the first books proposing psychology as a science—rather than a study of the soul, as philosophers and theologians saw it—began to be published.

And the great Charles Darwin himself added impetus to the notion that behavior was foremost a product of biology. Once started, Darwin seemed unstoppable. In 1871 and 1872 respectively, he'd published two books— *The Descent of Man* and *The Expression of Emotions in Man and Animals*— both of which explored comparable behaviors in humans and other species, suggesting an evolutionary explanation for the development of the brain and how it processed everything from logic to love.

It was, in its way, Darwin's answer to Wallace's proposal that an intelligent designer was required to explain the capabilities of the human mind. James's first course at Harvard—The Relationships between Physiology and Psychology—drew directly on the Darwinian view, exploring the ways that that chemistry and anatomy could explain the inner workings of the brain.

Although this new science of the mind fit into Tyndall's view of a mechanical universe—from the gas-fed engines of the stars to the chemically driven responses of living brain tissue—it prompted no contentious public debates, no warnings of a breakdown in cultural morals. Psychology's effect on nineteenth-century thinking was more insidious. By its very existence, it nudged theologians and philosophers aside as the primary students of human behavior.

James had established a small laboratory at Harvard to run psychology experiments. His work paralleled a similar, even more ambitious push in Europe, especially Germany, where scientists rushed to design and install equipment for monitoring physical evidence of psychological stimulus. Yet as this German approach, with its stress on measurement and quantifiable result, gained influence and acceptance, James found himself increasingly alienated from the science he had helped pioneer. A humanist at heart, and a contrarian, he gradually turned away from a mechanistic view of human behavior.

The new psychologists believed, as one leading American practitioner would say, that human behavior would be understood only when "all cells and fibres involved in each act of the mind or emotional state . . . could be numbered and weighed." James didn't quarrel with the importance of that principle; he'd written papers himself explaining that states of consciousness in the brains of animals resulted from chemical changes at the molecular level. He'd concluded that the same principle undoubtedly applied to the biology of human consciousness.

But he disagreed that accepting such results automatically defined him as a materialist and an atheist. James resisted the idea that along with its "brass instruments," science had acquired the ability to understand and explain everything about human thought and behavior. How did one measure questions about "the ultimate cause of existence" or whether life had meaning? Given that, James felt, science had no firmer grasp on Truth with a capital *T* than did the theologians. As for himself, the great questions were, he said, "hopelessly out of reach of my poor powers," just as they lay beyond the reach of his fellow academics.

James's desire to both explain the science and to look beyond it made his book on psychology intellectually challenging—and extremely slow going. Already his publisher, Henry Holt, was writing irritated letters demanding to know what was taking so long. And in addition to the textbook, James was now teaching philosophy. He was bothered by headaches, stressed by his drive to do so much so well. He began to obsess on his need for peace, quiet, even a change of scenery. After four years of living with a temperamental husband, Alice didn't hesitate. She encouraged him to take a break. He was so restless now, she said, she could more easily care for the boys in the company of her mother and sister.

WILLIAM'S FIRST STOP was at his brother's elegant London flat. Henry James Jr. was making a name for himself as a writer of style and substance. He'd published a rapid-fire sequence of well-received novels, *The American, Daisy Miller, Washington Square,* and *Portrait of a Lady*—all in a four-year span. Henry James had achieved a level of acclaim that his brother could only envy.

William relaxed into the visit. He spent afternoons in conversation at his brother's clubs, surrounded by an aromatic fog of tobacco smoke. He made occasional calls on scientists. He walked the sooty streets, enjoying Henry's company. Then he found himself suddenly alone. Back in America, Henry James Sr. was dying. Their mother had died of bronchitis earlier that year, and their sister, faced with this second impending death, felt overwhelmed. She asked Henry Jr. to come home.

William—the more high-maintenance brother—was to stay in England. "All insist William shall not come," his sister telegraphed. William debated returning home anyway, despite his nervous state, but had to admit that he probably wouldn't be an ideal deathbed companion. "How much better it will be to recollect him well than so decayed," he explained in a letter.

He wasn't certain that his father would even recognize him. Henry Sr. had suffered a series of small strokes after his wife's death. Paralyzed by illness and grief, the old man stubbornly turned his face to the wall and refused to eat. He died on December 18. At the end, he talked incessantly of a vision he kept seeing, of the stern faces of other old men, perched along the edge of a wall, watching as he passed by.

William, standing by in the smoky fog of London, learned of his father's death when he read it in the London *Standard*. He noted, with slight surprise, a burst of grief for his difficult parent—and an accompanying flourish of possibility. As he wrote to his wife, his father's death made him feel "as I never began to do before, the tremendousness of the idea of immortality. *If only he could be joined to mother.* One grows dizzy at the thought."

JAMES WAS WAITING for his brother's return, unenthusiastically wrestling with his psychology book, when he received an unexpected invitation from Edmund Gurney to dine with his philosophy club, "the Scratch Eight." Once a month, Gurney and seven of his friends—hence the group's name— met to eat and to argue over a different philosophical question.

As Gurney explained, he lived nearby and was an acquaintance of James's author brother. He politely presumed on that connection to invite

William to the December philosophy dinner. James enjoyed the evening—"I felt quite at home among them"—and his feelings were clearly reciprocated. He was invited to the next meeting of the Scratch Eight. But he especially enjoyed the company of Edmund Gurney.

His next letter home sang with enthusiasm about this new friend, "one of the first rate minds of the time, a magnificent Adonis, six feet four in height with an extremely handsome face, voice and general air of distinction about him, altogether the exact opposite of the classical idea of a philosopher."

Psychical research—and its implications—easily occupied their conversations, continuing in an exchange of letters after James returned to Cambridge in March of 1883. Gurney had accepted greater responsibilities at the SPR and was now honorary secretary. As he confessed to James, he still wondered at the strange enterprise, uncertain if the mysteries the group probed could ever be solved, or should be solved. He wondered, too, whether he was the right person for the quest.

"I doubt its compatibility, at any rate with my upsettable condition and easily fagged brain," Gurney admitted. And yet he was aware that the project needed someone of his stronger qualities—determination and intelligence—and who was financially able to devote himself to the work. There were hundreds of candidates for solicitor positions. The same could not be said of persons wishful of exploring the occult.

The choice, he knew, would require sacrifice. Gurney anticipated "loss of reputation, or rather (since I haven't much to lose) a gradual positive reputation for being weak in the head." Yet his doubts seemed petty when weighed against the possibilities.

In this moment in ontological history, with the aftershocks of the Darwinian earthquake still shuddering across the religious landscape, Gurney saw a chance to calm the tremors, to create a new, integrated worldview. If a scholar could connect science and faith, find the points where they met, a point where perhaps one approach might illuminate the other, Gurney believed that man might be able to make sense of life itself. He might even be able to define when life began—and when it ended. "Risks must be faced, whatever one does," he wrote, "and I feel no doubt the effort is worth the making."

Both the personal struggle and the determination to explore further

resonated with James. They were natural friends, he declared, and Gurney
agreed. In William James, he recognized a "rare and precious kinship, the
kind that made one think that Providence had done one a really good
turn."

"'Two lost souls!' you will say," James wrote later to a fellow philosopher, describing his bond with Gurney and his own growing interest in
psychical research. "But that is what remains to be seen."

GURNEY HAD AN IDEA about Mark Twain's dream vision. He and Myers
had been talking about what one might call the ordinary occult, experiences that unpredictably shadowed people's lives. They were interested in
dreams and premonitions as well as the haunts and spooks that reportedly
inhabited the waking world.

They proposed to treat such accounts—from Twain's dream to Catherine Crowe's tales of ghostly footsteps—as cumulative, as pieces of an elaborate jigsaw puzzle, broken and scattered, needing to be reassembled into a
picture of . . . well, they weren't sure what it would show, but Gurney and
Myers thought the image would somehow convey a greater truth.

As William James would emphasize, such stories arose in every book of
human history: "No matter where you open its pages, you find things recorded under the name of divinations, inspirations, demoniacal posses-
sions, apparitions, trances, ecstasies, miraculous healings and products of
disease." Supernatural events seemed to string together like a message, a
secretive code that so far remained undecipherable.

The SPR ran newspaper ads soliciting personal stories of encounters
with the otherworldly. Gurney spent hours each day sifting through them,
replying to personal accounts of apparitions, asking for confirmation, witnesses, documents. "I have been tremendously busy all winter," Gurney
wrote to James in early 1884, explaining his recent lack of friendly correspondence. By his own calculations, Gurney had written something like
sixteen hundred letters in the past two months, fifty-five already that day,
aside from his quick note to James, which was for pure pleasure: "I wish
you were not severed by the intractable Atlantic."

It was addictive work, despite the tedious clerical demands. Gurney
found himself almost unable to think of anything else:

One lives in a whirl of sporadic interests & small excitements—
whether A will answer this question satisfactorily, and B that,
whether C's mother really died the night he saw her appear at a
distance, or a night or two earlier, so that he might have heard
the news between—&c&c&c,

I find it difficult, almost impossible very often, to sit down & *read*
anything, & feel as if I was not improving but rather the reverse.
It is a bore that there are not more hours in the day, & more
Energy to be got out of one's "grey matter" between waking &
sleeping.

Even across the dark, glimmering distance of the ocean, James found
Gurney's enthusiasm infectious. Psychical research "is as worthy a specialty
as a man could take up," he wrote back encouragingly.

There was no assurance, of course, that Gurney would be able to as-
semble the puzzle into something connecting this world and the next. But
at least he was making an effort to study the pieces. James was increasingly
persuaded that for religion and its moral convictions to remain a center of
Western culture, its teachings needed a new foundation, built on both traditional theology and the newer realities of the scientific universe. He failed
to understand why more intellectuals—humanists and scientists alike—
weren't actively working toward that link, as Edmund Gurney was trying
to do.

Theologians and religious leaders, James wrote to a fellow philosopher,
mostly resisted the idea that faith, in light of modern science, should
include "belief in new physical facts and possibilities." Not all churches
could be judged as one, of course, but James found himself dissatisfied with
so many aspects of Victorian religiosity. The Catholics, steadfast in faith,
seemed to him determined to pretend that scientific discoveries were meaningless At the other end of the spectrum, however, he deplored the "blood-
less pallor" of the Unitarians' careful open-mindedness. "Of all the senseless
babble I have ever had occasion to read, the demonstrations of these philosophers who undertake to tell us all about the nature of God would be the
worst, if they were not surpassed by the still greater absurdities of the philosophers who try to prove that there is no God," he complained.

A few religious leaders, such as James McCosh, president of Princeton University, were attempting to integrate evolution into their teachings. McCosh proposed that the Darwin-Wallace theory served "to increase the wonder and mystery of the process of creation," and thus was a tribute to God's powers. Others were at least acknowledging the geological issues, the increasing evidence that life arose and changed gradually over millions of years. Most scholars now suggested that when the Old Testament prophet Moses talked of "days," he meant "ages." But many clergymen still rejected evolution outright, as the devil's path to a "might makes right" society, taking the position that since God had to be right, science had to be wrong.

As James wrote to a fellow philosopher that year, "I confess I rather despair of any popular religion of a philosophic character & I sometimes wonder whether th[e]re can be any popular religion raised on the ruins of the old Christianity?" He wished that his own community of science was working to reduce such hostility. Instead, researchers tended to imply, if not declare, that those who clung to Christian beliefs simply lacked the intelligence to understand the rational view of life. It seemed to James that scientists were missing an opportunity to be included in a discussion that could well shape the future. "Are the much despised 'spiritualists' and the Society for Psychical Research to be the chosen instruments of a new era of faith?" he asked. "It would surely be strange if they were, but if they are not, I see no other agency that can do the work."

At James's invitation, the charismatic William Barrett arrived in Boston in September 1884, en route to a science conference in Montreal. He was there to explain the psychical research being done in Europe—and to encourage the Americans to surpass it. With the blessing of the Sidgwick group, Barrett hoped to stimulate interest in a psychical research organization in the United States that would complement the British one.

Squeezed in among scientists, philosophers, and theologians from Harvard's Divinity School and area churches, Barrett described the research projects of the British Society for Psychical Research. He explained the SPR's goal—to explore those "remarkable phenomena, which are *prima facie* inexplicable on any generally recognized hypothesis" and which had yet to be credibly investigated.

He warned his audience to expect ridicule if they took up his challenge.

But that was true, Barrett, said of most new sciences. And yet given a chance, alchemy had developed into chemistry, and the star mappers of the past had become the astronomers of today. Barrett believed, and he would repeat this throughout his long career, that "sooner or later psychical research will demonstrate to the educated world, not only the existence of a soul in man, but also the existence of a soul in Nature," and he hoped that the Americans would also see the golden promise of that ambition.

LABORING IN CAMBRIDGE, Henry Sidgwick had formed one solid conclusion: that he was a terrible psychical researcher.

Everything seemed to flatten out when he appeared; knocks faltered, raps halted, spirits faded away. He always seemed to "paralyze the phenomena," he told his colleagues, and, depressingly, they agreed with him. He'd racked up hour after hour observing nothing happen. "I'm going to a haunted house," he wrote gloomily to Myers on a properly dark fall afternoon, rich with shadows. "Where I shall see no ghosts."

The fault belonged to him, Sidgwick wrote in his diary; he lacked the skills of his fellow investigators. Nora was more talented an observer, Myers more tireless, Gurney more acute in his judgment. No wonder they were doing more interesting work. But Sidgwick did have a few strengths he thought he could bring to the cause. He wrote them down too: He believed in justice. He was fair-minded. He was a good listener. And when he looked with satisfaction at his colleagues, Sidgwick had to note another talent: the ability to recruit excellent people. In point of fact, he'd recently persuaded another philosophy student to join their ranks, a cheerful cynic named Richard Hodgson, and he thought it might be one of his best actions to date.

Hodgson was a big, burly, vigorous man with a fresh, ruddy face and a shock of sandy brown hair. Born in Melbourne, Australia, in 1855, the son of a wool importer, he'd considered becoming a lawyer, graduating from the University of Melbourne with a law degree—but decided practicing law didn't seem intellectually stimulating enough. He decided, instead, on graduate work at St. John's College, Cambridge, choosing the school because his favorite poet, William Wordsworth, was an alumnus. Hodgson had read nothing better than Wordsworth's lyrical odes to nature, the poet's assertion that "One impulse from a vernal wood/may teach you more of man/of moral evil and of good/than all the sages can." He hoped, eventually, to hear and learn from nature as Wordsworth did. In fact, he scheduled time for it in his daily plans. Hodgson was—as Sidgwick had noted—ever disciplined in his habits, always determined in approaching his goals.

He got up at 7:30 every morning. He had breakfast at 8:00 a.m.: "one raw egg in one half pint milk, one slice of bread, one and a half cups of tea." He read till nine, worked on essays and letters until noon. He had bread and tea for lunch. He lounged and read "fiction or light poetry" till 3:45. He played lawn tennis until dinner. He went to a gym where he boxed three times a week, or fenced, or worked out with dumbbells. At 9:00 p.m. he had a supper of bread, eggs, and tea. He read poetry until bedtime, usually by 11:30. He allowed himself one cigarette before sleep. "Regularity for the organism is everything," he wrote his best friend in Australia, a fellow law student named James Hackett who had decided to remain there and become a lawyer.

Only nature and poetry, his two abiding loves, could cause Hodgson to slip a little from his standards of activity. "I do enjoy a leisurely walk home from lecture these days," Hodgson wrote to Jimmy Hackett. "The sunlight goes right through me." He read Wordsworth, Browning, Tennyson; he snuck in some hours to write his own doggerel verse, which betrayed a lurking, ironic sense of humor and a keen interest in ongoing debate over Darwinian evolution:

> *The Bishop knows better; for Huxley and Tyndall*
> *Have shown him that man goes away like a fly;*
> *And seeing the soul's such a regular swindle,*
> *He'll eat and he'll drink and tomorrow he'll die.*
> *We'll prove you and I by a laudanum potion*
> *The body is just what we always have thought her*
> *A middling arrangement of molecule motion*
> *Ammonia, and carbonic acid and water.*

Hodgson worried about Hackett's lack of mental stimulation, mired in the dull practice of the law. He promised to write his friend about all the interesting philosophical questions that came up at Cambridge and the personal ones. "Thy brain shall not be dormant while I live, O Jimmy!"

HODGSON KNEW he lacked the proper philosophy student's attitude, which he described as sitting at the professor's feet, and he knew he annoyed Sidgwick. "I was rather amused," Hodgson wrote home, "because he seems a goodly fellow and I thought I detected a feeling that he had better be rough with me with the intention of diminishing my confidence."

But in this, he underestimated his philosophy professor. Sidgwick saw in the young Australian much more than an inflated ego. He saw a smart, decent, hardworking man—and a natural investigator. He'd been looking for someone like Hodgson. He had a project in mind, the SPR's most ambitious investigation to date.

As had so many others, Richard Hodgson found himself persuaded by Henry Sidgwick. By December 1884, the young Australian scholar was in Bombay, India, on an expedition financed by his philosophy professor, in pursuit of an elusive but influential psychic.

On the surface, Helena Petrovna Blavatsky was an unlikely target. A fat, middle-aged Russian with a puglike face dominated by enormous, bulging eyes, she had the chain-smoker's habit of scattering herself and her listeners with cigarette ash. Yet Blavatsky carried such a reputation for mystical powers that she had been able to create her own religious group, the Theosophical Society.

Madame Blavatsky, as she liked to be called, had once lived in Tibet. She claimed to have a mystical connection to the godlike mahatmas of the Himalayas, who had the power of "astral projection"—sending their "astral bodies" to anywhere in the universe while their physical bodies stayed where they were. Her followers said that the mahatmas sometimes rose in a mist from her shoulders—or perhaps that was just the perpetual haze of cigarette smoke. Nevertheless, she could produce the most amazing physical effects. Shattered dishes mended themselves, apports materialized, and sealed letters, bearing personal spirit messages, wafted down from the ceiling of her Manhattan apartment.

Madame Blavatsky promised her followers a world of reincarnation, of balance with universe, and even—with time—of mystic abilities of their own. She had begun her Theosophical Society in New York, publishing a book of her philosophy called *Isis Unveiled* in 1877, but then moved her headquarters to Madras, India, to be "closer to the source" of her powers. She still spent much of her time in the United States and Europe, spreading her particular gospel. In the summer of 1884, she had attended an SPR dinner in Cambridge, where she charmed the company, despite her "unattractive appearance," Sidgwick noted. "She is a genuine being, with a vigorous nature both intellectual and emotional, and a real desire for the good of mankind," he wrote in his diary. He was impressed by her direct way of looking at people, and the down-to-earth cattiness of her conversation: "Thus in the midst of an account of the Mahatmas in Tibet, intended to give us an elevated view of these personages, she blurted out her candid impression that the chief Mahatma of all was the most utter dried up mummy that she ever saw."

Even Nora liked Blavatsky's direct manners. Further, the notion of astral projection was a fascinating one, and one that the society was eager to test. But—as happened so often with high-visibility psychics—Madame Blavatsky deftly avoided being tested. She explained that the source of her powers now lay in India. She'd built an elaborately gilded shrine in Madras, glittering with carved vines and blossoms, angels and animals, containing tiny drawers into which spirit letters would suddenly appear. It was unfortunate, she said, that the investigators couldn't work with her there.

Thus Richard Hodgson found himself in Bombay, admiring the shimmer of the fireflies from a hotel room paid for by Henry and Nora Sidgwick, and organizing a trip to the famous shrine of Madame Blavatsky.

"THE EVIDENCE PUBLISHED by the English society is of a nature not to be ignored by scientific men." So read the first circular of the American Society for Psychical Research, which was officially founded in early 1885, excited into being by William Barrett, three years after the creation of the British SPR. The American group, however, decided against being led by classical scholars, as the British society was. Its founding members, among them William James, determined to operate on purely scientific methods, to use only trained researchers as investigators. "Not that scientific men are necessarily better judges of all truth than others," James explained, but even he thought researchers tended to be more believable as experts. Scientists were better trained to gather evidence, "and what we want is not only truth, but evidence." There was a risk that the researchers' reputations might be tarnished by association with the occult—one had only to observe the effect on him, or Crookes, or Wallace—but, James noted, "how much easier to discredit literary men or clergymen!"

The ASPR remained open, of course, to members from other disciplines, and they had a respectable few, among them James's publisher, Henry Holt (still impatiently waiting for the book on psychology); Gardiner Hubbard, who would soon found his own organization, the National Geographic Society; the Asian scholar Crawford Toy, who had been expelled from Southern Baptist Seminary in Louisville for his efforts to reinterpret the Old Testament in the light of Darwinian science; Charles Everett, dean of Harvard's Divinity School; George Fullerton, a professor of philosophy at the University of Pennsylvania; and Minot Savage, a Boston pastor and author of *The Religion of Evolution.*

But the founders had chosen an astronomer, Simon Newcomb, who headed the Naval Almanac office, to stand as president, and other scientists, physicists and psychologists, to fill the remaining officer positions. James was especially pleased about Newcomb, whose opinions carried real influence. In addition to his naval appointment, Newcomb was a professor of astronomy at Johns Hopkins, revered for his painstaking mathematical recalculation of planetary orbits. He was the first president of the American Astronomical Society, and a past president of both the American Association for the Advancement of Science and the American Mathematical Society. If Newcomb set the right tone, "he will probably carry the others," James wrote, hopeful that more mainstream scientists would follow.

NEWCOMB'S TONE, as it turned out, was that of a man who had suddenly realized he had said yes when he meant no.

He began his presidency by writing an article for *Science* magazine in which he dismissed Barrett's work on mind reading and all subsequent experiments on the subject by the British society. Real scientists knew, Newcomb wrote, that thought was a mere bodily function, a product of internal chemistry and nervous system impulses, a self-contained product.

Newcomb himself had yet to read a single piece of convincing evidence that a thought could be "transferred" from one mind to the other.

Why was it, Newcomb demanded, that if thought reading existed, and presumably had existed for as long as had human thought, "no living person knows any more about the conditions of transference today than men did a thousand years ago?" Why was it that after all his experiments, William Crookes had been forced to go to physicists like Barrett in search of a working theory of action? Why was it that neither chemist nor physicist had been able to provide such a theory?

True, Barrett and his colleagues had produced results suggesting that thought transference did exist, in fact if not in theory. But Newcomb found that easy enough to dismiss. He suspected their findings were simply lucky coincidence, overstated by sloppy research methods. None of the British society's work impressed Newcomb. He considered Nora Sidgwick's plan to study reports of haunted houses no more than the process of collecting "very scientific children's ghost stories." And he was dismayed that Gurney and Myers had posted ads in the British papers, asking ordinary citizens to write and describe any supernatural experiences. Newcomb didn't consider that a scientific way to gather information. It made their goal far too public, he scolded. They should have been doing all their data collection by means of private inquiry.

William James had told the new ASPR members that psychical research was "bringing science and the occult together in England and America." But speaking for science, and speaking as the ASPR's president, Newcomb hadn't seen a fact or a study to convince him of that.

Oh, really? replied Edmund Gurney, who was angry enough to fire off a rebuttal letter to *Science*. Didn't it make scientific sense to gather a large number of answers and then sift out the silly ones? It was very well to insist only on private inquiry, if one had an unlimited budget and a decade or two to do the work. "It is worrying to think of the stores of authentic evidence which are untapped & which will gradually become unavailable. I believe that the speed with which the subject gets on will depend more than anything else, on the number of cases collected, sifted and classified. It is the *un*scientific portion of the community that our progress really depends on."

For all his scientific knowledge, Newcomb seemed unable to understand the research requirements or respect the process of psychical research, Gurney wrote to James, adding that in his opinion the ASPR's new president had "managed to get considerably off the rails." Even more, Gurney and his colleagues were dismayed by Newcomb's frontal attack on thought transference, especially since the British association had made it a goal to see those experiments become accepted science. Myers had even coined a new name for the phenomenon—*telepathy,* from the Greek *tele,* "far," and *pathy,* "feelings." The existing term, *thought transference,* Myers said, was too limited; their experiments tended to show that feelings, sensations, and images also could be conveyed mind-to-mind. He defined telepathy as a transmission, evoking the telegraph and the telephone, summoning the concept of mental messages humming across some as-yet-to-be-discovered strands of energy. Telepathy was fundamental to all supernatural events, Myers said; it could explain premonitions, sympathetic reactions to another's emotions. Telepathic communication might also prove the best scientific explanation for prayer, at least prayers that seemed to be answered.

Not that the experiments yet came near any such exalted conclusions. But they did consistently suggest that person-to-person communication occurred outside the obvious channels. Myers and Gurney themselves had conducted a neat little series of tests, exploring the ability to transmit a sensation such as taste. The tests were modeled after William Barrett's suggestion that hypnosis might make a person a better "receiver," more able to concentrate on reception.

As Gurney described them, some of his more tantalizing results came from studies with two young men, identified only as A, who was put in a hypnotic trance, and B. Both subjects were concealed from each other by a screen. Gurney made no effort to prepare either man for the experiments. He simply brought a package into the test room. The box contained some small vials, each packed with a strong-tasting material. He administered the substances, one at a time, to his two subjects. The result was a sometimes-comical series of exchanges.

"I suddenly and silently gave [B] some salt, motioned to him to put it in his mouth. He did so; and [A] instantly and loudly exclaimed 'What's this salt stuff?'" Gurney then provided sugar. His hypnotized subject sighed with relief, "Sweeter, not so bad as before." When Gurney once again gave B a spoonful of salt, A was exasperated: "I told you I liked sweet things, not *salt*—such a mixture."

It was the preciseness of these experiments that bothered Newcomb. They seemed to him too good to be true. That went not just for Gurney's taste tests, but other SPR experiments as well. He especially disliked a series of tests in which one participant was asked to draw a picture and then mentally transmit it. A second participant, sitting blindfolded, was asked to draw the image "received."

The original drawings were kept simple: the outline of a goblet, the wooden frame of a chair, a fatheaded fish. So were the companion drawings, those sketched out by the receivers. But they were still recognizably fish or furniture, only a little more crooked than the original work. That was what Newcomb found so unlikely, that a person blindfolded could draw so relatively accurately. Therefore he suspected cheating.

By this time, William James had grown as exasperated with Newcomb as Gurney was. James also wrote to *Science,* attacking Newcomb's comments on the SPR: "To brand as dupes and enthusiasts a set of gentlemen as careful as these English investigators have proved to be, seems to me singularly unjust."

And then he dismantled Newcomb's criticisms, almost point by point. Despite Newcomb's new ASPR title, James stressed, the gentleman had no experience in these matters. He had not done any observation himself, but merely complained about the work of others. Further, if Newcomb had bothered to test his misgiving about the drawings, he would have seen how feeble it was. James had made a point of checking his own ability to draw blindfolded. With a cloth over his eyes, he found that he could sketch outlines of fish and fruit as well, if not better, than the ones Newcomb criticized. To prove it, James sent some of his drawings to both the journal and Newcomb.

The journal's editors had printed Newcomb's text with an admiring editorial note concerning the astronomer's "acute observation" of psychical research. James characterized him instead as a "critic without substance." The one thing everyone could agree on was that this was not a good beginning for the pro-science approach to psychical research.

. . .

"The whole thing is a fraud," Hodgson wrote to his friend Jimmy Hackett in March 1885, after four months hunting down his quarry, who had now reputedly left India under an assumed name. "I seal and register this letter because part of Madame Blavatsky's performance consists in getting hold of letters, by bribing or otherwise, steaming them open, getting knowledge of contents and reclosing." She also amended letters once they'd been opened. Hodgson admitted to a kind of reluctant admiration for her complete nerve: "She is a tolerable imitator of other people's handwriting, will stick at nothing, and is one of those remarkable clever unprincipled women one reads about but seldom meets."

His report to the Sidgwicks, also sent sealed and registered, provided the incriminating details at length. Hodgson interviewed witnesses to Blavatsky's séances. He'd compared letters from the mahatmas with examples of her handwriting. He'd hired handwriting experts to analyze opened letters. He had asked to inspect the shrine and its miraculous cabinet, and when refused the first time, had turned up time and time again, so tirelessly that he'd finally simply worn her guardians out.

Once in, Hodgson had patted and pulled the shrine's ornate walls until he'd discovered that some of the drawers were double-sided, opening into a chamber behind the gold facade, which turned out to be Madame. Blavatsky's bedroom. He'd then managed to obtain a confession from her servants that they'd passed letters through the panels into the carved drawers on the other side.

Hodgson had scarcely left the building before it mysteriously burned to the ground, turning its secrets into ashes. He'd no doubt she'd ordered the destruction of evidence. Hodgson sent all the letters he'd collected to Nora Sidgwick, who reanalyzed the writing and was also impressed by the quality of the forgeries, not to mention the quantity.

The SPR's report on Madame Blavatsky minced no words: "For our own part, we regard her as neither the mouthpiece of hidden seers nor as a mere vulgar adventuress; we think that she has achieved a title to permanent remembrance as one of the most accomplished and interesting imposters in history."

In its way, Hodgson's report gave the British SPR a perfect retort to those who characterized them as gullible dupes. It was hard to imagine even the most accredited scientist doing a better job of demolishing a medium's reputation.

WHAT WAS THE WORST, the most maddening, and the most annoying aspect of being a psychical researcher? It was "this perpetual association in the eyes of the world with 'intellectual whoredom,'" Edmund Gurney wrote to William James in the spring of 1885.

Gurney was only repeating a phrase used by the ever-waspish John Tyndall, who was still making it his business to discredit psychical research whenever the opportunity appeared. That attitude, that inflexible hostility, was "the real rub—the thing that occasionally makes the work so trying & galling." Gurney was tired of Tyndall and his ilk assuming that he and the other SPR members *enjoyed* listening to stories of Himalayan gods who mailed letters to paying clients, that they thrilled to chasing down "humbug" and con artists, that somehow "this is what one likes. . . . But I did not mean to give vent to cursing & swearing when I began this letter."

He did mean to pass along good news. Nora Sidgwick had finished her detailed, coolheaded analysis of ghost stories, almost four hundred in total. And he and Myers had begun sifting through the hundreds of responses they'd gotten through advertising for personal accounts of visions and dreams.

As Gurney told James, "Our plan has been to spread the nets very wide, get a big haul, & then viciously sift. Perhaps 5 per cent of the cases sent or heard of bona fide prove worth something. But we should never have got the 5 if we had not laid ourselves open to the 95." What kind of story fell into that golden 5 percent? One of Gurney's favorites came from a country squire in eastern England.

In the mid-1870s, the squire had for a neighbor a young man who owned a farm that ran adjacent to his lands. They occasionally hunted together, occasionally had a drink, but the older and younger man did not build what one would call a strong friendship.

On a chilly March evening in 1876 the squire met his neighbor walking across the corner of his estate, and they fell into conversation, mostly somber, about the low prices of farm produce. As they walked back toward home, the young man invited the squire to "come and smoke a cigar."

This neighbor had rarely been so hospitable, and the older man was a little surprised. He had to refuse, though, explaining that he had a dinner engagement. Yet the other man seemed unusually anxious about the rejection, repeating the invitation several times, before saying good-bye.

About ten that night, after he'd returned home again, the squire went into his library to retrieve a book he was reading on the natural history of birds. He had barely settled into an armchair, near the window, when he heard the front gate open and shut, and a hurry of footsteps approaching the house. The stone walls of his house suddenly seemed transparently thin; he could hear the visitor's labored breathing. And then a scream, a wail of horror, fading into sobs of agony.

"Of my fright and horror I can say nothing—increased tenfold when I walked into the dining room and found my wife sitting quietly at her work close to the window." The squire's wife had heard nothing, but looking up, catching a glimpse of her husband's pale face, she said, immediately, "What's the matter?"

"Only someone outside," he replied.

"Why not go out and look, as you always do when there's a strange noise?" she inquired.

"There is something so queer and dreadful about the noise," he answered, and he remained with her, in that comforting circle of domesticity.

He hadn't been willing to step into the dark just then.

In the morning, though, the squire went out to look for footprints. The ground and paths were frosted with cold. They sparkled in the sun. The glittering surfaces were unmarked, except by the light tracks of a passing hare. There were no marks where he had heard the visitor approach. He couldn't understand it.

Only a few hours later, a horrified friend came by to tell them that their neighbor had committed suicide by drinking a glass of prussic acid, a cyanide potion distilled from a dye called Prussian blue. The young farmer had bought the poison that morning, telling the chemist that he needed to kill a dangerous dog.

It had been waiting for him at home, after he said good night to the squire. Prussic acid poisoning was a terrible death, rackingly painful. The body looked as if the young man had died screaming. The county coroner thought perhaps the death had occurred after ten or so, the previous night.

As the squire wrote to Gurney, he had no explanation for what happened. He knew only that it had. And just this one time, he emphasized, never before or since. He wanted Gurney to know that he wasn't a sensitive man, always hearing things and flinching at shadows.

Gurney checked out the story anyway, compiling statements from the man's wife, neighbors, the chemist who'd sold the poison. He'd verified even the claims of frosty weather by reading newspaper accounts of the time. It was only then that he put the squire's story into the credible 5 percent pile. Slowly the pile was growing, though. And in the slim stack of stories there, Gurney thought he was just beginning to see the hint of an outline, the faint form of something that might be real.

"I think our case is really strong enough to show that the subject ought to be earnestly prosecuted," Gurney told James. He did agree with Newcomb on one main point, though: They needed to be absolutely, completely clear in their work and their findings: "I feel that every sentence written on these matters ought to absolutely *reek* of candour."

Nora Sidgwick was fairly impervious to insult. But she could have wished that Simon Newcomb had actually read her analysis of ghost stories before deriding it as child's play.

Her report had, actually, been so skeptical as to infuriate the British society's most dedicated spiritualists and believers. Nora had flatly declared that most spooky tales were spun like sugar, thready creations of foggy nights and fevered imaginations, "of such a nature as to justify the contempt with which scientific men generally regard" such reports. Some angry members of the SPR had threatened to quit over her account; as her husband noted with affection, she was completely untroubled by that reaction.

As Nora saw it, she needed to eliminate illusions and hallucinations and then to decipher the meaning of the few credible stories. For instance, almost everyone who claimed to see a ghost described the dead person as fully dressed. Why should that be? Why should there be "ghosts of clothes," as Nora put it? One might argue that a ghost represented a dead person's spirit or spiritual energy, but it was difficult to accept that shirts and skirts also passed into an afterlife. Why would their wardrobe return with them? Why—as Fred Myers somewhat sarcastically said—should the theory of metaphysics encompass "meta-trousers and meta-coats"?

On the surface, Nora believed that clothes-wearing ghosts were a point against the stories being true. They made no scientific sense. On the other hand, Nora thought that if she could figure out why so many credible witnesses saw them, then she and her colleagues might move closer to understanding why people saw ghosts at all.

Because the one thing she did believe was that many of the people she talked to did see ghosts—or at least were convinced of the fact. "I can only say that having made every effort—as my paper will, I hope, have shown—to exercise a reasonable skepticism, I yet do not feel equal to the degree of unbelief in human testimony necessary to avoid accepting at least provisionally the conclusion that there are, in a certain sense, haunted houses." If one accepted that conclusion, she said, then one also needed to ask what, precisely, created that "certain sense" of being haunted.

WILLIAM JAMES WOULD remember that summer night, the one that found him alone, thinking about a dead child, as darkness barely lit by a "clouded moon."

He stood on a Cambridge street, looking up at a bedroom window in a house he had once occupied. The night was quiet and the window was shuttered, tucked under the eaves of the second floor, shadowed by memory.

His third son, Herman, had been born in that room, in January 1884, a child so irresistibly chubby and cheerful that he'd immediately needed a less serious name. James nicknamed him Humster. And now it was August 1885 and Humster had been dead for more than a month, buried in a wicker cradle basket near his grandfather's grave, marking yet another month in a terrible year.

James's wife, Alice, had been quarantined with scarlet fever in the early spring. For three months the children—six-year-old Harry, three-year-old Billy, and one-year-old Herman—lived with Alice's mother in her nearby Boston home. Even after Alice began to recover, even after they washed the bedroom, disinfected it with sulfur fumes, stripped the wallpaper, and repainted the walls, James and his wife worried about allowing the children back home.

But their littlest boy begged to stay, clung to his mother on visits. Finally James and his sister-in-law, Margaret, removed Billy and Harry to New Hampshire and allowed the baby of the family to stay with his mother. It was a decision made with love, a decision that went rapidly wrong. Alice, still fragile, developed whooping cough. Herman caught the infection, which turned rapidly into a vicious pneumonia. In answer to his wife's terrified letter, James rushed home to find his youngest son dying, racked by fever and convulsions.

Herman died in his mother's bed on July 9. Two days later, they buried him under a small pine tree in the family plot in Cambridge Cemetery. William and Alice wrapped a little white flannel blanket around their son's casket, and when it was lowered into the ground, they surrounded it with flowers and leaves. James confessed later to one of his aunts that he'd always looked down on such rituals. "But there is usefully a human need embodied in any old human custom and we both felt this." They left their son cradled in wicker, smothered in branches and leaves, and "there he lies."

And, yet, on this night in late August, James walked back to their old rental house just to stand there in the hazy moonlight. He mourned the brief flutter of his son's existence. On the following day, James wrote to a cousin, "It *must* be now that he is reserved for some still better chance," some promise beyond life on Earth.

He had no intention of trying to prove a very personal wish; no plan to consult a medium on behalf of his son. That he ended up doing both, William James would always consider a strange and remarkable coincidence.

. . .

LEONORA EVELINA PIPER was twenty-six years old in 1885. The wife of a Boston shopkeeper, she was slightly chubby, neatly dressed, her light brown hair caught carefully up into middle-class respectability. The Pipers *were* middle-class respectable. Leonora, her husband William, and their one-year-old daughter, Alta, lived with his parents in a tidy house in the Beacon Hill neighborhood.

But the neighbors whispered that young Mrs. Piper wasn't quite as ordinary as all that. She could tell people things about their lives that she couldn't have known. Sometimes she told them family secrets that they didn't know themselves. The rumor was that she could hear the voices of the dead.

According to Leonora's parents, the first hint of any such ability occurred during her childhood in Nashua, New Hampshire. At the age of eight, while playing in the garden, Leonora felt a sudden, sharp blow on her right ear and heard a sudden sibilant hiss. The child stood shocked as the snakelike sound slowly resolved itself into an S, then the name Sara, then a sentence.

Screaming, she ran into the house, calling for her mother, holding the side of her head. At first her mother could get no sense from the hysterical girl. Finally, the child stammered, "Oh, I don't know! Something hit me on the ear and Aunt Sara said she wasn't dead but with you still." She was so upset that she scared her mother, who wrote up the incident, the day, and the time in her diary that night. Several days later, they received a letter from the aunt's husband, telling them that she had died, on the day, about the time that hissing voice had spoken into the child's ear.

Young Leonora (then Symonds) and her family wanted nothing to do with any of it—the whispering voices or the whispering neighbors. There were children celebrated for psychic gifts; the notorious Fox sisters came to mind. The Symonds family had no intention of seeing Leonora become such a freak. They put the eerie little moment behind them and raised their daughter as an upright member of the Methodist Church. She married William Piper when she was twenty-two, and if it hadn't been for a troubling illness, she might have left it at that, a moment of otherworldly fright in a country garden.

From her sixteenth year, Leonora had walked with a slight limp, the result of an ice-sledding accident. Another child's sled had crashed into her on a snowy hill, damaging a knee and, more seriously, causing internal abdominal bleeding. In the years after, she'd been conscious of a dull ache across her midsection, and now, after the birth of her first child, the pain grew sharper.

Frustrated by the inability of doctors to diagnose the cause, she visited a clairvoyant, an elderly blind man who claimed that he could contact spirits to aid in healing. When the psychic touched her, she grew almost immediately dizzy. "His face seemed to become smaller and smaller," she said, and to the shock of the other sitters and the psychic himself, she tumbled to the floor.

Voices were ringing in her head. She could hear only one of them clearly. She gathered herself up, went directly to a table, scribbled a note, and handed it to an elderly gentleman waiting his turn with the psychic. The gentleman, a Cambridge judge, said it was a message from his dead son, "the most remarkable I ever received." She went back several more times to the psychic's parlor, but she found she was becoming the attraction.

Strangers were now coming to the Pipers' home and asking Leonora to go into a trance for them. Alarmed, she retreated. She didn't want to be a medium. She was expecting a second child. She wanted to be a mother and a respectable wife. Still, she had to wonder if this was some God-given gift. Leonora Piper prayed over it. She couldn't quite bring herself to turn away all the callers. In the late summer of 1885, she let a friend talk her into sitting with a Boston widow.

The widow was Eliza Gibbens, the mother-in-law of William James.

As James recalled it, some two months after Herman's death his mother-in-law came to visit, fizzing with excitement and disbelief. The young Beacon Hill medium had told her about family members, both names and facts, "the knowledge of which on her part was incomprehensible without supernormal powers." It was so impossible that Mrs. Gibbens determined to investigate further. She sent her daughter Margaret to visit Mrs. Piper the following day with a tougher test, a letter in a sealed envelope.

Don't open it, Margaret said to Mrs. Piper, just tell me something about the person who wrote it.

Reading sealed letters was an easy trick for mediums of the time. They could conceal an alcohol-soaked sponge in a hand or sleeve and surreptitiously soak the paper with it, rendering it transparent—and decipherable—until the alcohol evaporated. They had only to briefly distract the visitor until they could return the envelope and reveal its contents. With a good distraction, most mediums also showed a flair for opening and resealing envelope flaps in time to avoid detection.

But Mrs. Piper kept things simple that day. She held the letter in front of her. And then she slowly described the writer—where she lived, why she had moved across the Atlantic. Even if she had somehow been able to sneak a look at the letter, Margaret had deliberately chosen from a correspondence written only in Italian, which Mrs. Piper definitely did not know. Margaret Gibbens and her mother decided to tell Alice about their find. She was still so thin and pale after the whooping cough and Herman's death; like William, she had found it difficult to let the little boy go. Perhaps this would intrigue her, cheer her up a little, perhaps she could ask this odd medium about her lost son.

"I remember playing the *esprit fort* on that occasion before my feminine relatives," James wrote later, "and seeking to explain by simple considerations the marvelous character of the facts which they brought back." He considered himself something of an expert on psychic performances. He and the Reverend Minot Savage, of the ASPR, had been visiting the more notable mediums of Boston, meticulously attending séance after séance, and learning lessons in what both men considered to be brazen fraud. "This did not, however, prevent me from going myself, a few days later, in company with my wife, to get a direct personal impression."

Mrs. Piper met them in the front parlor of her in-laws' home, offering the couple seats in a pair of stiff wingback chairs. They had not given her their names, and to James's relief, his mother-in-law and sister-in-law had earlier refused to disclose their identities.

He'd emphasized to Alice that she must follow strict psychical research rules. They wouldn't mention any connection with the earlier visits. They wouldn't provide any information about their family at all. They wouldn't ask leading questions. They wouldn't answer such questions, either. They would listen politely and, he predicted, be bored senseless until they returned home for dinner.

Leonora Piper settled herself into a fatly stuffed armchair, leaning back into a nest of pillows. They began talking about the weather. It had been an unusually gentle autumn. Late-afternoon sunlight glazed the room. Her eyes began to drift shut. Her head turned sideways against the pillows; a faint tracing of goose bumps rose on her skin. She would always describe the sensation of slipping into a trance as like descending into a dense and chilly fog.

Her voice seemed to deepen a little. Mrs. Piper began repeating the names she had given to Alice's mother and sister. And then she began fumbling for other names, mumbling them, getting them not quite right. "The names came with difficulty and were only gradually made perfect. My wife's father's name of Gibbens was pronounced first as Niblin, then Giblin," before the right name was fumbled out. It was as if she couldn't pronounce the words at first, or couldn't quite hear them right.

As Mrs. Piper added details to the names, James, as he later wrote a friend, became increasingly uneasy. It could be that the young psychic knew everyone in his wife's family on sight. She could be incredibly lucky in guessing about the domestic life of strangers and their relatives. Or it could be that most improbable, scientifically impossible conclusion—that this woman "was possessed of supernormal powers."

Before coming out of her trance, Mrs. Piper did ask about a dead child. But that too could be an easy guess. Many couples had lost children to illness. He watched the medium's entranced face, her closed eyes, and the slight frown between them. It was a boy, she said, a small one. Herrin? Herrin? No, she would finally conclude the boy's name sounded more like Herman.

"PEOPLE WHO FLY into rages are such a bore," Nora Sidgwick remarked to her husband in the fall of 1885. "I really think the spiritualists had better go."

The British psychical research society maintained a policy of allowing all interested parties to join. The member list included some prominent mediums and some of their more devout spiritualist followers. These members professed to agree with the plan for skeptical research. But it now appeared that they hadn't meant the word skepticism to be taken quite so literally.

Many spiritualists remained angry over the perceived negative findings in Nora's analysis of ghost stories. Some quit in outrage over the exposé of

Madame Blavatsky. The mediums in the organization were infuriated by Henry Sidgwick's distaste for professional practices. Most of them would no longer even speak to him. Sidgwick had learned from reading the spiritualist newspaper *Light*, that its editors were conducting an angry crusade against an *Encyclopaedia Britannica* decision to have Nora write the article on spiritualism. "And we had fondly thought they would be pleased!" he noted in his diary. They had assumed that spiritualists would admire dedicated researchers taking an interest in the occult, sorting out the legitimate phenomena from the fraudulent. Instead, it seemed that the churches of spiritualism were not so different from the churches of Christianity. To his perception, neither could tolerate evidence contradicting what they wished to believe.

It wasn't just Nora who was antagonizing the membership. Richard Hodgson was back in England, cheerfully adding fuel to this already smoldering sense of resentment. This time, Hodgson wasn't investigating a specific medium but a specific practice. He had decided to take apart the practice of slate writing.

Hodgson liked the direct approach. For this occasion, he'd persuaded a young conjurer to begin holding slate-writing séances, following the principles employed by some of London's more acclaimed mediums. The conjurer in question, S. J. Davey, was a frail man, slight and bookish. Plagued by ill health, he had learned to wile away his resting hours by practicing magic tricks. An SPR member who had visited a number of mediums, Davey had been depressed to recognize his own well-practiced conjuring methods enlivening the séances. He was delighted to team up with Hodgson, who had a very specific plan in mind. It involved Davey as performer and Hodgson himself as "manager."

As they set it up, Hodgson would schedule performances, inviting sitters to witness the amazing talents of his friend. Davey would do the rest. As a team, Hodgson and Davey held more than twenty memorable "séances" in which Davey caused messages to appear on the surface of locked slates, the answers given with amazing precision. When Hodgson wrote, "What is the specific gravity of platinum?" the responding message was a scientifically accurate snap: "We don't know the specific gravity."

Hodgson and Davey were explicit from the beginning. All of this was accomplished through stage magic. The "platinum" message had involved a rather neat exchange of locked slates while Hodgson was inspecting a table from below. There were no spirits in these séances, merely fast hands, hidden devices, distraction—and a wish by those participating to believe in magic. Following hard after the SPR's earlier exposés, this struck the spiritualists as a gratuitous attack. They did indeed fly into the kind of accusatory rage that Nora Sidgwick found so tiresome.

They also refused to accept that the Hodgson-managed séances weren't real events. Davey was too good; he must be a renegade medium, maliciously seeking to discredit his own profession. Even Alfred Russel Wallace was drawn into the fight, taking the side of the spiritualists and making it obvious how far he had drifted from the Darwinian mainstream for the moment. "Unless all can be explained," Wallace declared in late 1885, "many of us will be confirmed in our belief that Mr. Davey was really a medium as well as a conjurer, and that in imputing all his performances to 'trick' he [Davey] was deceiving the Society and the public."

ACROSS THAT INTRACTABLE ATLANTIC, the psychical researchers suffered through no such dramas, but then, their best work was being done in careful secrecy. Afraid that he was somehow being gulled, James had asked the Reverend Minot Savage, his partner in other medium investigations, to take his own look at Leonora Piper.

Savage paid an anonymous visit to Mrs. Piper, telling her only that he'd heard of her trances and wished to observe one. At Savage's first sitting, Mrs. Piper talked about his father, who had died years before in Maine. "He calls you Judson," she said. Judson was Savage's middle name. Only his father and his half-brother, also dead, had ever called him that.

Savage wrote up a detailed, somewhat incredulous description of the sitting: "She went on to say, 'Here is somebody else besides your father. It is your brother, no, your half-brother, and he says his name is John.'" Savage noted that his half-brother's name was John, which was not an uncommon name and could easily have been guessed. That wasn't the part that shocked him. It was what came next, the way Mrs. Piper went on to describe with painful accuracy, partly in pantomime and partly by speech, the method of his half-brother's death. She finishing by saying, "When he was dying, how he did want to see his mother."

Savage's half-brother had never lived in Boston. He had died in Michigan, two years earlier, "in precisely the way the medium had described the facts." John was a "mother's boy," Savage said; he'd died calling out to her. Aside from the details, the way Mrs. Piper talked in trance, as if these dead people were somehow sitting beside her, was unnerving, to say the least. Still, both Savage and James had learned that trusting any medium was a risky business. Perhaps Mrs. Piper had some kind of detective bureau at her service; perhaps she'd somehow discovered both men's extended family history. Savage decided to give her a test that even the best spy network would find hard to beat, a complicated, multilayered kind of experiment.

To start, Savage told his daughter to visit Mrs. Piper, using an assumed name. His daughter asked a friend to write the letter seeking the appointment, so that not even the handwriting would be a clue. Another friend then provided three locks of hair, placed in a book—front, middle, and back—so as to keep them separate. Savage's daughter knew nothing about where the hair had come from, or even whether the donors were still alive.

As she told her father, she hoped the visit wouldn't be too ridiculous. Beyond that, she was keeping her expectations low.

Minot Savage wanted more than just to present Mrs. Piper with a tricky test. He also wanted to evaluate a concept that currently intrigued psychical researchers. Called "psychometry," it was the idea that material objects could convey information to people, or conversely, that some people had the ability to read information from rocks and wood, paper and metal.

The word *psychometry* had been coined in 1842 from the Greek words *psyche,* "soul," and *metron,* "measure." But the idea itself was woven through folklore from many cultures and many, many years past. Generations of ghost stories derived from the belief that a building could contain memories of a murder, that terror could inhabit a place for years to come.

Most academics regarded the idea as so rank with superstition that it was hardly worth discussing. About thirty years earlier, though, a Boston geologist had decided to run psychometry tests on his wife—who claimed some psychic ability—and a few of her friends. His experimental approach was simple in the extreme, using the materials he knew best. The geologist wrapped rocks in paper packets and asked the sensitives to tell him something about the contents. A fragment of lava from the Hawaiian volcano Kilauea elicited the response, "It seems as if an ocean of fire were pouring over a precipice"; a limestone pebble with glacial scratches: "I am going, going, and there is something above and around me. It must be ice. I am frozen in it."

As critics pointed out, his test subjects could have been guessing while they tried to feel the rock through paper. And even if there was something to psychometry, no one could explain how an inanimate object could communicate with a person. The concept provoked questions rather than theories: Was there an energy left by intense events? Could a house somehow retain the "memory" of a violent murder; could a rock recollect the savagery of an eruption? Or was it all just a fascinating dream of supernatural abilities, kept alive by the occasional coincidence?

Savage didn't know himself. But he did know that Leonora Piper had shown a possible psychometric talent in that Italian letter test given her by Margaret Gibbens. It was that which led him to devise the experiment with the locks of hair. Like his daughter, he kept his expectations as low as possible—which meant that both of them were shocked by the outcome.

As Savage wrote in his report to the ASPR, "After Mrs. Piper had gone into a trance, these locks of hair were placed in her hand, one after another. She told all about them, gave the names, the name of the friend who had asked her to bring them, and even asked why one person had cut the hair from the end, where it was lifeless, instead of nearer the head.

"My daughter, of course, did not know whether any of the names given or statements made were correct or not. She made notes, however, and found that Mrs. Piper had been accurate in every particular." James and Savage collaborated on a formal report to the American society, which admitted that both men found Mrs. Piper inexplicable. They urged that the ASPR invest serious time and money investigating all the questions they had raised about the medium. Here, they said, was a rare opportunity to put "real" science to work, and they hoped not to lose it.

5

m

INFINITE
RATIONALITY

$\mathcal{S}$UDDENLY, in a shock of glorious insight, Edmund Gurney realized the truth about immortality. The moment was breathtaking, luminous. He alone had solved the infinite puzzle, seen the path to eternal life.

Unfortunately, as he complained to William James, that gleaming certainty melted like ice in a thaw—about the time the nitrous oxide he'd received at the dentist's office wore off. "I was only waiting for breath & energy to inform the lucky dentist, when it somehow lapsed." Still, even leaving the office, Gurney had felt sure that he could write a brilliant journal article. By the time he got home, the brilliance had worn off, but Gurney still thought he might write James a letter filled with useful insights. "Then the evening came & somehow the letter did not seem as if it would have much in it and didn't get written."

Oh, those transient moments of narcotic genius. He'd had them before; he suspected he'd have them again, given his particular habits. Gurney suffered from occasional bouts of neuralgia, sharp bursts of pain exploding along the nerves in his face. He'd taken to relieving the pain by inhaling chloroform or, during particularly miserable episodes, laudanum, a solution of opium and alcohol.

There was no stigma to it; self-medication was a stylish fact of the times, prescribed by doctors, practiced by the educated and the wealthy. William James also used chloroform to treat neuralgia, and he had once published an essay on the fleeting brilliance of nitrous oxide inhalation. Fred Myers and Dick Hodgson had both experimented with hashish, although Myers had simply fallen asleep, and Hodgson had disliked the giddy loss of control: "I'm not born to be hashish'd out of my organism." Doctors widely prescribed laudanum for pain, stress, depression, and menstrual symptoms. In Europe, medical researchers experimented with cocaine as a therapeutic agent. In 1884 a young Austrian psychiatrist, Sigmund Freud, had published a well-received research paper, "On Cocaine," based partly on his own use of the drug as a stimulant and counterdepressant.

Insights imparted by narcotics, as James had ruefully written, always proved mere illusion—even if one could remember them. Once he carefully jotted down every thought that occurred to him under the influence of nitrous oxide. In the morning he had pages with single words written over and over—God, day, night, prayer—scribblings that "to the sober reader seem meaningless drivel but which, at the moment of transcribing, were fused in the fire of infinite rationality."

Consistently, sadly, by the dry light of morning, by the time one returned from the dentist, there was nothing left but a tantalizing memory, ashes with not a phoenix in sight. Gurney regretted those ashes. If only there were such a fire that burned longer, with the sustained light to illuminate the hidden promise of his project.

GURNEY AND MYERS were building their studies of the everyday occult into a book. They'd given it the title *Phantasms of the Living* as a way of describing the supernatural events that they deemed credible—that 5 percent of all reports Gurney had mentioned to James. They wanted both to recount those stories and to put them in perspective, reinforcing their "evidence" with theories that might explain it and arguments to support those theories. They wanted the book to be so good, so compelling, that it would sway the world.

The intensity of that desire, the demands of the project, were so overwhelming that they'd recruited a third collaborator, thirty-year-old Oxford graduate Frank Podmore. Thin, perfectionistic, ever serious in outlook, Podmore had won honors in both classics and natural science at Oxford with his meticulous scholarship. He now supported himself as a postal inspector, which left him time for other interests. And he boasted a Hodgson-like flair for investigation that Gurney thought could only strengthen the evidence they would accumulate. The only problem was in making sure that Podmore and Myers didn't quarrel throughout the entire project.

The two men—one obsessed with details, the other's eyes on the philosophical horizon—had been at odds almost since the work began. It wasn't just the book, although that most concerned Gurney at the moment. Sidgwick had also asked Podmore and Myers to investigate the mystery-cloaked life of Daniel Dunglas Home. They were arguing over that as well.

The reclusive psychic Home had finally lost his long battle with tuberculosis. At fifty-three years of age, he had died in southern France on June 21, 1886. His Russian wife buried him in a Paris cemetery under a tombstone inscribed simply, cryptically, "To another discerning of Spirits."

Home's career still haunted the field of spiritualism. As Podmore would write, most professional mediums were "imposters of a sufficiently common place type," easy to catch in fraud and chicanery. The Davenport Brothers, Anna Eva Fay, Henry Slade, Florence Cook—to name a few on a very long list—all had been exposed with very little effort.

No one could say that of Home. There were whispers of exposures at private sittings, but never a public unmasking. His cheats, if that is what they had been, went undetected. Even Robert Browning, who had so deftly skewered Home as "Mr. Sludge," confessed to Myers and Podmore that he had never been able to catch Home in fraud. Browning only wished, as so many other critics had done, that he had been so lucky. As Podmore wrote, "commonplace is the last epithet that could be justly applied to Daniel Dunglas Home."

Like many before them, Myers and Podmore could not agree on how to define D. D. Home. Myers was inclined to consider Home a rare talent, a medium with unusual gifts for tapping into occult forces. Podmore was inclined to consider him an unusually deft conjurer, one of the greatest hypnotists of his time, or some combination of the two.

If Home could induce mass hallucinations, that might explain some of his more improbable effects—floating out a third-floor window into the

London night, summoning shadows into shapes, elongating his body so that he appeared to stretch as thinly as a bit of India rubber. "Probably some of the more marvelous feats described at Home's séances can be analyzed into sensory deceptions of this nature," Podmore would eventually conclude.

Hypnosis would not serve to explain the carefully catalogued experiments that William Crookes had conducted. Podmore could only speculate that in such cases Home cheated, and that scientific procedures were insufficient to detect such a sophisticated manipulator at work. But that left the depressing conclusion that one unusually deft medium could easily outwit some of the best scientists of the day. "In Home and his doings all the problems of spiritualism are posed in their acutest form," Podmore would write ruefully. "With the marvels wrought through him or by him the main defenses of Spiritualism must stand or fall."

Meanwhile, *Phantasms of the Living* was nearing completion. The authors and their friends debated every aspect of it, including the title, which seemed adequate rather than engaging. Still, they couldn't decide on a better one, and "phantasm," meaning illusion or hallucination, seemed an objective way to describe what Gurney, Myers, and Podmore had collected: reports of phantoms and ghostly voices, premonitions and warning dreams, flickers of connection between the living and the dead, and the implications of what lay behind them.

It was the implications that worried Henry Sidgwick, or at least, how to present those controversial ideas. The plan had been for Myers and Gurney to share equal billing as lead authors. But Sidgwick began to think that a poor decision. True, Myers was the more graceful, more literary writer. But Gurney wrote with more simplicity and thus greater clarity. More important, Sidgwick admired Gurney's methodical approach; he was less easily carried away by an idea than was Myers. Given the hostility simmering around them—from scientists and now spiritualists—Sidgwick suspected that one strong voice rather than "two heads on one neck" would best serve the project. He wanted Gurney's level head to represent the project. The sensitive Myers, however, would need careful handling.

Sidgwick set the stage over dinner. For the discussion after the repast, he made sure that a soothing fire warmed his library, which he stocked with plenty of good brandy. As the snifters drained, Sidgwick proposed that they turn the book over to the more cautious Gurney. To his diary, Sidgwick admitted it had probably been a lost cause from the beginning. Even the finest brandy could not prevent Myers from seeing what was afoot. "I was palpably aiming at ousting F.M. and leaving E.G. sole author; estimating the superior trustworthiness of the one in scientific reasoning more important than [the other's] literary superiority."

Other people might underestimate Sidgwick, mistake his gentleness for passivity. His friends knew that he was relentlessly stubborn. When the evening was done, Sidgwick had worn his colleagues to a compromise. Myers would write the introduction, and Gurney the bulk of the book. "I could tell that M. was annoyed but he handled it admirably."

Not entirely admirably. Some of Myers's frustrations and resentments spilled into his writing, which took on a sharp, challenging tone. The provocative question that he chose to confront first was that of proof: In a modern age, Myers asked, could religion endure based on faith alone? Could any God—Christian or otherwise—survive in an age where religion feared science and science denied faith?

It was into that divide that Myers saw psychical research bravely marching. The goal was to bridge research and religion, to show that they were not incompatible, that one could even help explain the other. Yet there was as much peril as promise in such an effort. Orthodox science might contradict religious dogma, as Darwinian theory had appeared to do, but its proponents had in general—aside from a few aggressive hard-liners such as Tyndall—left matters of the spirit to those who professed to understand such things. The faithful could dismiss theological contradictions. They could declare evolution and natural selection irrelevant to the contemplation of God and His powers. Those who didn't like the conclusions of geologists, biologists, and physicists could ignore, dismiss, even ridicule.

Psychical research, however, explored the very mysteries that other scientists had, thus far, eschewed, and that the faithful reserved as their own. If a fossil were erroneously identified, a rock poorly dated, scientists could be embarrassed, shamed, forced to make corrections. But the error posed no threat to theology. If research failed entirely in its goals, Myers said, if chemistry and physics and astronomy all together stumbled, the faithful would in no way suffer. But if he and his colleagues failed, if they could find no evidence of an afterlife, no proof of otherworldly powers, they might further undermine the church's promises of immortality. And there, indeed, was the risk: he and his friends might find the evidence for "the independence of man's spiritual nature and its persistence after death"—or they might confirm the existence of nothing, no promise beyond that of a quiet grave.

If dedicated psychical researchers could detect no flicker of noncorporeal spirit out there in the universe, then the final result, Myers thought, would be to prove Tyndall and his allies right. Perhaps the truth was that religion couldn't stand up to science. Perhaps all the belief systems—from Christianity to Hinduism—existed only because they had been created in a time before scientific challenge.

The very argument gave Sidgwick the willies. The traditional churches had not rallied to psychical research; clergymen tended to point out disdainfully that the immortality promised in the Bible in no way resembled the spooks and specters of SPR investigations. Sidgwick had hopes that, with time, the churches would alter that stance, accept that miracles needed evidence, that science and faith could stand together in the pulpit. He worried that Myers's confrontational position might be so alienating as to make such discussions impossible. He feared his colleague's forthrightness would further isolate their cause.

"M. says roundly to the Theologian, 'If the results of our investigation are rejected, they must inevitably carry your miracles along with them,'" he scrawled anxiously in his diary. In the privacy of his study he confessed that his colleague was probably right. The quest for proof defined much of their efforts and of spiritualism in general. No wonder that people embraced the idea of physical proof, even to the ridiculous appearances of levitating furniture and materialized spirits. The rational man or woman needed some substance to his saints, Sidgwick agreed, with one caveat: "This is, I doubt not, true, but is it wise to say it?"

IF ONE DOUBTED the demand for physical proof, one needed only to watch the latest craze in spiritualism, "the talking board" phenomenon, sweep across the United States, catching everyone from workingmen to children in its fascination.

"Planchette is simply nowhere," declared the *New York Tribune,* compared to this newest tool for talking with the dead.

Rectangles of shellacked and polished wood, talking boards had the words *yes* and *no* printed in the upper right and left corners. The entire alphabet was inscribed across the center. In the lower corners were the words *Good Eve* (for "good evening") and *Good Bye.* The board was simple to use, relying on a planchettelike device that could slide across the surface, its point indicating answers. Users would sit on either side of the board, their fingers resting on the pointer, and watch for it to move, pointing to "yes" or "no" or using the letters to spell out more complicated sentences.

"I know of whole communities that are wild over the 'talking board,' as some of them call it," one man told the *Tribune* reporter, and the eerie stories they told about its powers contained "things that seem to pass all human comprehension or explanation."

Talking boards worked faster than spirit raps or table tiltings. They were easier to read, by far, than the scribbles of a pencil-wielding planchette. And they were so popular that they were soon being mass-marketed by companies including Sears Roebuck, who determined that they needed a more commercial name. By the late 1880s, the boards would be rechristened Ouija boards (or "yes-yes" boards, from the French *oui* and the German *ja.*)

The boards' success, their apparent proof of spirit communication, also drove their popularity: "One gentleman of my acquaintance told me that he got a communication about a title to some property from his dead brother, which was of great value to him," an excited user told the *Tribune.* "Attempts are made to verify statements that are made about living persons, and in some instances they have succeeded so well as to make the inquirers still more awe-stricken."

GURNEY WAS SO LOST in his search, so consumed by his determination to make others see what lay before him, that he had almost forgotten about his pretty wife and sweet young daughter.

Kate Gurney cared nothing for the metaphysical debates and theological questions in which her husband had buried himself alive. She liked parties, gossip, and lively conversation. Edmund seemed such fun when she married him; now he had become so very serious and, really, a bore. He encouraged her, of course, to enjoy society without him. If there was any advantage to her situation, it was that her handsome husband's chronic inattention and habitual absence gained her a bit of sympathy. Even Myers had heard the sad plight of Kate Gurney discussed in elegant drawing rooms.

Phantasms of the Living sprawled through the Gurneys' home. Papers overflowed Edmund's study, spilling out onto furniture and the floor. It lay in packets, in piles, in stacks and drifts, so that the elegant Mayfair dwelling seemed at times submerged in *Phantasms,* just as Edmund's life had become submerged. When the physicist Oliver Lodge, one of the SPR's telepathy researchers, came to visit, he found himself rather alarmed at the state of the Gurneys' paper-strewn home, the disorienting sensation of swimming through a vast ocean of documents. He wondered if *Phantasms of the Living* was worth its cost in time and reputation: "Attention to such gruesome tales seemed to me a futile occupation for a cultivated man."

Surrounded by his documents, his fair hair disordered, his blue eyes shining, Gurney convinced Lodge that such concerns were unnecessary. His home might resemble chaos, but a pattern was emerging from the flood, an answer floating just under the dark surface of this river of sorrow.

The emerging pattern derived from number and repetition, in the way that so many stories of the dead echoed each other, by their weird consistency. As mystery piled on mystery, strangeness upon strangeness, each ghost story—based on diaries from the past and letters in the present—gained power from each other:

A British nobleman was traveling with friends in Sweden. They decided to make a long day's journey toward Norway and finally halted, exhausted, at a coaching inn.

It was one o'clock in the morning when they stopped, an icy winter night. The nobleman was so thoroughly chilled that once he had stumbled into his room, his teeth still chattered in his head. He decided to see if a steaming bath would thaw him out.

"While lying in it and enjoying the comfort of the heat, after the late freezing I had undergone, I turned my head around, looking toward the chair on which I deposited my clothes, as I was about to get out of the bath."

Sitting on the chair was a man he'd once known well. They'd been the closest of friends as university students together. But his friend had joined the Civil Service and gone to work in India; it had been many years since they'd spoken.

The nobleman was so startled that he stood up, slipped, and went sprawling onto the floor. When he pulled himself up, his friend was gone.

It was such a strange moment that he couldn't bring himself to tell anyone. But he wrote it down, noting the date in his diary—December 19, 1799. And he wondered what it meant; he worried about his friend—and himself.

"No doubt," he wrote to Gurney, "I had fallen asleep and the appearance, presented so distinctly to my eyes was a dream." Except that he could not quite persuade himself of that. He had never had a dream that felt quite so vivid, quite so compelling, and quite so awake.

And except that shortly after his return home, he received a letter from India, announcing his friend's death and giving the date as December 19, 1799.

The woman woke suddenly in the night: "I felt uneasy, and sat up in bed." She looked around the room and saw one of her sons, Joseph, standing at the door.

His head was heavily bandaged and his face bruised and bloody, especially about the eyes. He was wearing a dirty white nightdress. He stood silent, looking at her "with great earnestness," and then seemed to fade away.

At breakfast she told her family—four daughters gathered round the table—that she was prepared for bad news about Joseph. They laughed it away: "It was only a dream and all nonsense." But within days, in the midst of that frozen January of 1856, she received the news; her son had been killed in a steamer collision on the Mississippi, during which a mast had split apart and fallen onto his head.

Another brother, who'd rushed to the accident, had been there when Joseph died, calling for his mother. His head had been wrapped with bandages, as it "was nearly cut in two by the blow and his face dreadfully disfigured." Joseph had died wearing a white night-dress, soiled with blood and dirt.

The time of his death, almost to the minute, matched the moment when his mother had seen him watching her from the doorway.

The theater was crowded in Toronto. The merchant was taking a night off during a trip from England, relaxing in the company of a Canadian businessman.

He was looking down from the dress circle where they sat when suddenly a shadow, or a flicker, in the pit below caught his attention. He leaned over and saw a man standing below, looking up at him. He leaned further, and "I recognized in the features my twin brother, who at that time was in China."

His brother stood half in shadow, half in an oddly golden light that made his features startlingly clear. "I instantly exclaimed to my friend, 'Good God! There is my brother!' pointing at the same time to the figure."

"I cannot see anyone looking up here," said his friend, peering over the railing. But the merchant was so excited that he rushed down into the pit, calling for his brother. There was no one standing, no one looking up, no one resembling his brother in the crowd around him. "I am not superstitious, nor a spiritualist, but could not get over the startling circumstances for some time."

When he returned to England, he learned that his brother had died at the French Hospital in Shanghai on October 6, 1867. It was the night the merchant had been at the theater in Toronto.

The woman was sitting in her mother's bedroom when her seven-year-old nephew came running, tumbling, into the room.

The boy was pale and breathless: "Oh, Auntie, I have just seen my father walking around my bed."

She replied, "Nonsense, you must have been dreaming." The child's father was traveling in another country, she reminded him, on business far from home.

But the child would not be comforted. He refused to return to his room. So his aunt tucked him into the bed with her. An hour or so later, she rolled over, glanced idly toward the fireplace, and saw her brother sitting in a chair by the fire.

"What particularly struck me was the pallor of his face." Her nephew was asleep. "I was so frightened, knowing that at this time my brother was in Hong Kong, that I put my head under the bedclothes." But she heard his voice, calling her name. Once, twice, three times, and then it faded away.

In the morning, she told her mother, who advised her to make a note of it. She carefully wrote it down, 10:00 p.m. on August 21, 1869.

When the next mail arrived from China, there was the letter telling of her brother's death from heatstroke on August 21, 1869.

A British clergyman was taking a summer evening walk over the downs near Marlcombe Hill. He was composing in his head a congratulatory letter to a good friend whose birthday would be two days later, on August 20, 1874.

He had barely begun when a voice spoke sharply in his ear: "What, write to a dead man; write to a dead man?"

The clergyman turned hastily around, expecting to see someone behind him. There was only the fading light glazing the grasses with gold. "Treating the matter as an illusion, I went on with my composition." The same voice spoke again, this time louder and with some impatience: "What, write to a dead man; write to a dead man?"

Again, he turned around. Again, there was no one there. But now he was afraid that it wasn't an illusion.

After hurrying home, he wrote the letter and sent it anyway. "In reply [I] received from Mrs. W the sad, but to me not unexpected, intelligence that her husband was dead."

The man jolted upright in bed. It was four o'clock in the morning. Someone had just gripped his hand. The touch was as cold and thin as water.

He exclaimed to his wife, startled by the feel of those chilly fingers.

He caught a glimpse of a woman leaving; there'd been something about the way she moved, the set of her dark head, that had reminded him of his aunt.

But the man and his wife were in Nottingham, and on this early June morning in 1880, the aunt was supposed to be on a steamer heading for the United States.

He leapt up to check the front door of the house. It was on the chain. He returned, saying to his wife that he feared his aunt was dead.

"You're dreaming," she replied. Her diagnosis was that he'd eaten too large a supper before going to bed.

Two weeks later, they received a letter from his aunt's solicitor. She had died at sea on the day that he'd felt that ice-water hand in the middle of a summer night.

Gurney called the stories "crisis apparitions" because of their critical timing—the voice, the touch, the shape, all seemed to appear close to a moment of extreme injury or death. Out of thousands of responses, he'd kept only seven hundred for the book, only those with some tangible evidence behind them—diary notations, conversations with others at the time of the event, death certificates, newspaper coverage and obituaries, letters of corroboration, details such as the theater program from the businessman's evening at the opera to support his story and its timing. The investigators conducted personal interviews. They calculated time differences between place of death and place of vision. The compiled data accounted for most of the piles of paper that now inundated Gurney's house and his life.

None of these spirits were the chain-clanking, blood-spattered, terrifying ghosts of fiction. Sidgwick, who enjoyed telling a good ghost story—once enthralling a roomful of children with his harrowing account of "The Bloodthirsty Bluebells"—often complained that real ghosts were never so much fun as the gore-streaked phantoms that appeared in novels. Even Gurney admitted that reading up on crisis apparitions was "far more likely to provoke sleep in the course of perusal than banish it afterwards."

The apparitions' power was in their very repetitiveness, the similarities of "unlooked for detail," the consistent way the visions appeared and faded, the person after person telling the tale, the documents and the double checks. Everyone had heard the occasional similar tale once or twice. But as he and Myers told their colleagues, it was hard to explain "the effect on the mind of a sudden, large accumulation of direct, well-attested and harmonious detail."

As Gurney insisted to Lodge, the stories might be monotonous, but they added up. He had no doubt about that. The question was—added up to what?

"IT IS NOT from professional mediums—so numerous in the United States—from slate writing 'materializations' and kindred performances, that we can look for any enlightenment whatever on the positive side in the course of psychical research."

So wrote Richard Hodgson, in the summer of 1886, in the summation of his formal report on slate writing, detailing the elegant cons that he had run with S. J. Davey. Hodgson emphasized that he and his partner had merely copied the craft of working mediums. "I may conclude with a warning which I wish to give, especially to our members in America, viz: that nearly all professional mediums are a gang of vulgar tricksters who are more or less in league with one another."

The members of the American Society of Psychical Research needed no such reminder. Their annual report on mediums—barring the curious anomaly of Leonora Piper—had degenerated into a list of exposures of professional practitioners, seven in Boston alone within one year. "Such a state of things hardly tends to encourage [the] committee in the active pursuit of this class of phenomena," the report stated.

The American scientists had been dismantling other spiritualist claims as well, such as a popular theory that sensitives gained their power from an unusual ability to detect magnetic signals. Joseph Jastrow of the University of Wisconsin led that investigation. Jastrow began with a simple demonstration. He used a dynamo to charge a large magnet, generating a magnetic field. He then asked self-anointed psychics, sitting in an adjacent room, to tell him when the field was strong and when it was not. The first experiments yielded impressively positive correlations, leading him to wonder if some of the participants could, indeed, "feel" the magnetic pulse.

But gradually, Jastrow became aware of another possibility. He himself could occasionally hear the rumble of the dynamo and a low clicking that sounded with demagnetization. So he surrounded the dynamo and magnet with soundproofing insulation and then repeated and repeated the experiment—exhibiting a dogged, Henry Sidgwick–like patience. In the following 1,950 tests of the sound-muffled magnet on ten mediums, all the results were negative. Or, as Jastrow wrote: "We conclude then that our experiments, as far as they go, fail to reveal any sensibility for a magnetic field." His conclusion was that these professional mediums had no special talent, making them at worst liars and cheats, at best victims of a mental illness that caused self-delusion.

The ASPR's committee on haunted houses and apparitions had reached a similar conclusion regarding mental illness. Unlike Gurney and his colleagues, they saw no pattern of reality in haunted moments. They perceived instead a pattern of insanity or instability. Even the strongest mind could falter, the ASPR researchers pointed out; the perception of a ghost could be no more than a moment of mental fragility.

A general diagnosis of mental illness was, in fact, the first conclusion that a significant number of the American scientists liked. It suggested, at least, that psychical research might be of some use in the study of human psychology, that belief in the supernatural might be used as a tool, helping to identify aberrant brains. It allowed the researchers to quit dabbling in the occult, make their way back toward more respectable pursuits. It suggested that the marriage of scientists and psychics had been ill conceived.

William James gloomily mailed the report to a friend, calling it "a rather sorry 'exhibit' from the 'President's' address down." The society seemed to prefer attacking evidence of the supernatural to exploring it. "I suspect it [ASPR] will die by the new year."

DID THE PSYCHICAL research community plan to do anything beyond debunking mediums and discrediting supernatural claims? That was Alfred Russel Wallace's rather plaintive question. Could a positive result be seen in their future?

For all his interest in spiritualism, Wallace had remained one of the more innovative and productive scientific thinkers of his time. In the years since he'd published his first paper on natural selection, he helped pioneer the study of color vision and its development. He'd researched alpine corridors, glaciations, and island environments, describing what would come to be known as the Wallace effect of reproductive isolation. He'd also practiced what he preached about scientists upholding ethical standards for the betterment of society. In 1885 Wallace had supported a minimum wage for workers and proposed the unheard-of idea that manufacturers should be required to label goods, specifying their contents.

His fellow scientists fell short of such scrupulosity, he thought— especially those investigating spiritualism. While traveling in the United States on an 1886 lecture tour, he'd become particularly irate with the ASPR's apparent insistence on exposing every medium that passed before them. He was even angry with William James, who, despite his qualified support of Leonora Piper, had helped discredit some of the more popular physical mediums in Boston.

It seemed to Wallace that the scientists were missing the point of parlor séances and the bedazzlements in which professional mediums dealt. He didn't deny that floating furniture and slate writing were trivial. But he thought such demonstrations were "the only means of compelling attention to the subject, and this is more particularly the case with those imbued with the teachings of modern science."

Wallace complained that scientists were impossible to please. If they observed physical effects, they characterized them as fraud. If they were asked to study telepathy or other mental demonstrations, they responded that it wasn't worthwhile, since there was no measurable physical effect. He knew that even the SPR leaders considered him too gullible. Oliver Lodge said that the self-educated Wallace lacked the objectivity he might have gained from formal scientific schooling. More charitably, Fred Myers thought Wallace too fundamentally decent to comprehend the base dishonesty of the medium trade. "There are natures . . . which stand so far removed from the meaner temptations of humanity that those gifted at birth can no more enter into the true mind of a cheat than I can enter into the true mind of a chimpanzee."

But Wallace took the position that his critics should first accomplish something. "I think your constant allegation of fraud on mere suspicion unreasonable & unscientific," he wrote James, angry at the latter's dismissal of some of Wallace's favored mediums. "You ask for facts & proofs on our side but offer only suspicions on yours."

THE TWO FAT volumes of *Phantasms of the Living* were at the printer. Its publication date was set for October 1886. Not soon enough. Too soon. Everyone concerned suffered nervous jitters.

Sidgwick worried that no one would read it. Then he worried that people would read it and "select the weak stories to make fun of." Then he worried that such selections would diminish the book's influence. "It will have one advantage—hard to get in these days—that there never has been a book of the kind."

Gurney worried over the volume's imperfections, "the monotonous assortment that calls itself a cumulative proof. . . . It is a ragged affair. Things that have taken days of dull inquiry will look no better than if they had been picked up in 3 minutes." Still, he wrote to James, "to be sincere I am glad to have done with it, & feel sure that the time, being *my* time could not have been better spent. And life, which has been for many months a petty fever, will now assume its more normal ditch-water aspect."

Myers worried that they would fail to make people *think*. Despite Sidgwick's cautions and admonitions, he'd maintained his challenge to organized religion. But he had shifted his primary attack. Religious dogma was irritatingly, willfully blind, Myers thought. But scientific dogma was worse. He sought to counter efforts by traditional researchers to marginalize the work of psychical research. "We conceive ourselves to be working (however imperfectly) in the main track of discovery," he wrote, pointing out that the SPR's investigations had already repelled "the crazy wonder-mongers." The difference was that he and his friends did not choose to limit scientific scope—or to define reality—within the narrow boundaries that the "ruthless hand of science" allowed. They hoped to persuade the research community to be more intellectually adventurous and "to lay the foundation-stone of a study which will loom large in the approaching age."

Science's tight rein on reality, Myers said, reduced the universe to a large machine and people to small ones. Scientists declared human free will to be an illusion, and emotions like love to be vestigial instincts. "Our vaunted personality itself is seen to depend on a shifting and unstable synergy of a number of nervous centres, the defect of a portion of which may alter our character altogether." Research was stripping people of complexity and the world of promise and reward, he argued. "The emotional creed of educated men is becoming divorced from their scientific creed."

Given that scientists did not yet know *everything*—at least, so Myers believed—he deemed it far too soon to declare questions of immortality and spirit off limits to rational men. If scientific leaders were to be honest, they would acknowledge that they didn't hold a monopoly on the important questions of human existence. Rather than discouraging those questions, they would seek to help answer them.

As well as issuing a sweeping challenge, though, Myers could also offer some very specific ways in which researchers might join the investigation. The SPR offered some conclusions and some theories on which to build:

1. Telepathy, by which Myers meant the transfer of thoughts and feelings from one mind to another—was a fact in Nature.
2. Phantasms, by which he meant impressions, voices, or figures of the dead and dying, were seen by their friends and relatives with a frequency beyond chance.
3. Telepathy might explain these phantasms, since clearly they represented action of one mind on another. The "second thesis therefore confirms, and is confirmed by the first."

To borrow a historical analogy, Myers compared launching a new science to geographical exploration at its most world-changing. He invoked the explorers Magellan and Columbus, feeling their way through unknown waters, "ploughing through some strange ocean where beds of entangling seaweed cumber the trackless way."

The seaweed itself might "foreshadow a land unknown"; the peculiar patterns of crisis apparitions might lead to a far more detailed understanding of communication at time of death. If Columbus could stumble into America by such a way, adventurous researchers too might blunder their way into another world.

.   .   .

IN SUPPORT OF Myers's call to action, Gurney cited the best evidence so far assembled for both telepathy and crisis apparitions. In support of thought transference, he reviewed all the work; from Barrett's early studies to some rather neat new work he'd done himself on the telepathic transmission of sensations, such as taste.

Gurney had set up a testing lab in which one taster and two "recipients" each sat in a different room. The sender in one room tasted what he or she was given, with no foreknowledge of what it would be. Immediately after, each recipient in his or her individual rooms was asked to describe it. Gurney simply listed some of the results.

Powdered nutmeg
Response from two different recipients:
• Ginger
• Nutmeg

Powder of dry celery
• A bitter herb
• Something like chamomile

Worcestershire sauce
• Something sweet, also acid, a curious taste
• Is it vinegar?

Bitter aloes
• Something frightfully hot
• It is a very horrid taste

Gurney proposed that shared perceptions came from a transfer of mental energy. He likened it to the transferred vibrations from one tuning fork to another. He didn't know yet—nobody did—how that transfer occurred or what form of energy might carry a thought from one person to another.

The mind was so clearly an unreliable instrument—not predictable, not consistent—that it tended to complicate all efforts to make sense of telepathy. Sometimes the "transfer" was pitch-perfect, sometimes it was nonexistent. Sometimes it was as if they were measuring one thought in two heads; sometimes it was as if the experimenters were testing two different species on two different planets. Some people seemed adept at sending and clumsy at receiving, some the opposite. Some had no talent at all for the exercise. Some exhibited an almost terrifying awareness of the thoughts in someone else's head. Some people called nutmeg ginger. Worcestershire sauce became vinegar. Some of their drawing examples were close copies, some borderline. A sender drew a profile of a man with a beard; the recipient drew a man's profile, but beardless. A downward arrow turned into a shooting star. The name C-L-A-R-A became C-L-A-R-V.

What mechanical system, what physical method of transmission, Gurney wondered, could possibly explain such a wildly varied range of results? His speculation was that the solution lay in the slightly wrong answer. Perhaps the information might be sent in one form, but it could then be altered—bent even—by the mind of the receiver. A man might try conveying the taste of Worcestershire sauce to a woman. The woman might have always been unusually sensitive to the vinegar that was part of the sauce ingredients. He sends a complex taste; she keys into one aspect, and Worcestershire sauce becomes vinegar.

It was, he thought, like a thousand everyday conversations: Henry and Nora Sidgwick talking about their garden, her mind calling up images of glorious roses and starry lilies, his mind flinching from memories of pollen-dusted air and hay fever miseries. The same subject filtered through different experiences—even between two people who knew each other well. Why should anyone expect the sharing of a thought to be easy or predictable?

The wonderfully complicated ways that two minds might interact also came to intrigue—and even obsess—Gurney's colleague Myers. Pondering the way the mind works below its conscious level, the way it adjusts and personally tunes the information it receives, Myers would come to believe that conscious thoughts and responses are influenced by information stored in the "subliminal mind," a concept that would later be called the subconscious.

The more detail-oriented Gurney focused for the moment on how such mental exchanges might explain puzzles in telepathy—and in crisis apparitions as well. His idea was, as Myers had proposed earlier, that crisis apparitions might be created telepathically. In a last burst of energy, dying or desperately ill persons might send their thoughts flying toward a friend, a family member, someone held in their mind during their final moments.

Blazing with intensity, those thoughts might reach the minds of people normally impervious to telepathic energy. And in the mind of the receiver, that personal contact might be also be altered, transformed by memory and emotion into a voice in the night, an image, the touch of a hand. His theory, thus, might well provide the answer to Nora Sidgwick's earlier objection to "the ghosts of clothes."

Why do we see apparitions as clothed, when cloth is clearly inanimate, a material without any chance of an afterlife? The most likely answer, Gurney thought, is that our own mind puts clothes on the ghost. We receive an impression of our dying mother; that image becomes wrapped in memories, tangled in them, dressed by them. She appears in her favorite Sunday outfit, feathered bonnet and all. Or our father appears clothed as he always was, on the way to the office; watch chain gleaming across the gray wool of his waistcoat.

Gurney's favorite of many examples that he saw favoring this interpretation came from a London woman recalling a vision in August 1884.

The lady was out for a drive in an open landau when she saw an old friend walking down the street. She was startled because she had thought her friend was vacationing by the seashore. She was further surprised that her friend wore a favorite sealskin jacket. She knew the warm jacket well, but it seemed odd attire in the hazy warmth of a summer day.

She called out a greeting to her friend. To her additional surprise, the woman did not answer; indeed, her head stayed slightly turned away. "For the next 10 minutes or so, I was puzzling to think what could have brought her back to London," and why she had behaved so. The woman wished she had asked her coachman to turn around so that she could have caught up with her friend. When she got home, she called the butler to inquire if she had a visitor and was told no. She wondered if her friend had gone to her sister's home instead and sent a servant to find out—but she was not there either. Three days later, she saw in the London paper that her friend had died at the seashore, on the day that she saw her in the street.

The story had the same elements of so many other crisis apparitions, but, for purpose of the argument, Gurney wanted to focus on that unlikely sealskin coat. He suspected the detail had come right out of his correspondent's mind, that she had received a "telepathic impression" of her friend and filled it out, adorning her friend with that familiar fur. In other words, he said, it is our own mind that creates the look of the apparition, adds to it out of our own experience: "One percipient may hear his parent's voice; another may imagine the touch of his hand upon his head; a third may see him in his wonted dress and aspect; a fourth may see him as he might appear when dying . . . others may invest the disturbing idea with every sort of visible symbolism, derived from their mind's habitual furniture and their wonted trains of thought."

The traditional scientist would counter, Gurney knew, that every apparition could be explained as coincidence. Dreams and daydreams, images of a friend or relative, frequently flitted through everyone's consciousness. Such images might randomly occur at the time of a crisis, and if so, they would be far better remembered than those lacking such dramatic narrative. The dreams and the voices would take on an undeserved sense of mystery and power, as one might argue about Twain's dream of his dead brother.

"The question for us now," Gurney said, "is whether these coincidences can, or cannot, be explained as accidental. If they *can,* then the theory of telepathy—so far as applied to apparitions—falls to the ground." But if he could prove, with statistics for instance, that "the same sort of startling coincidence is again and again repeated," far beyond what reasonable chance would predict, then he thought he could counter the objection.

Gurney's other name for crisis apparitions was "hallucinations of the sane." Almost everyone interviewed described these events as one time only. Almost everyone was uncomfortable or unhappy with the fact that they'd seen a dead woman while driving in the street, heard a deceased friend's voice. Most of those interviewed insisted that they were not superstitious, loathed spiritualism, and were baffled by the experience.

The question that needed to be answered, then, was how often did rational people, in the course of their everyday lives, find themselves caught up in such a hallucination? Clearly some kind of survey was needed, a census of hallucinations, which was exactly what Gurney proposed. He had done some preliminary sampling, randomly mailing out question-naires and receiving 5,705 answers to the question, "Since January 1, 1874, have you—when in good health, free from anxiety and completely awake—had a vivid impression of seeing or being touched by a human being, or of hearing a voice or sound which suggested a human presence, when no one was there? Yes or no?"

Twenty-three persons during those twelve years had reported a "visual hallucination." This might seem small, Gurney said, but based on the adult population of Britain and the number of deaths, he calculated the random chance of such a vision was perhaps a trillion to one. That made 23 positive responses out of 5,705 people seem a rather startling preliminary result. Emphasis on preliminary; Gurney thought only a much larger census, with perhaps ten times as many people, would eliminate statistical doubt.

The survey, inadequate though it was, also yielded another interesting correlation. All of the visual hallucinations occurred within twelve hours of the death of the person seen. If Gurney also analyzed the 702 crisis appari-tions recounted in *Phantasms of the Living,* more than half of them, 401 to be exact, appeared near the moment of death, and another 25 occurred toward the end of a fatal illness.

Questions and obstacles jostled for room in his thoughts. There were cases when a group of people reported seeing a crisis apparition. Did a dying person really have the energy to send out multiple messages? And if haunted houses were ever proved to exist, it would be hard, not to say impossible, to argue that a momentary flash of connection could account for a ghostly presence seen by many people, over and over again, for many years. For that, one would have to prove psychometry, the idea that the traumatic energy of a death could permeate an object or a place indefi-nitely. The intrinsic weirdness of the work seemed an obstacle in itself.

"The peculiarity of the subject cannot be gainsaid and must be boldly faced," Gurney wrote. "For aught I can tell, the hundreds of instances may have to be made thousands. If the phenomena cannot be commanded at will, the stricter must be the search for them; if they are exceptionally tran-sient and elusive, the greater is the importance of strong contemporary evi-

dence." But even without that, he could feel himself slipping into certainty—that peculiarity was simply characteristic of the way that the dead
said their good-byes.

PHANTASMS OF THE LIVING'S 1886 publication placed it in the midst of
another tumultuous, hectic, forward-thinking year. In Paris, Louis Pasteur
founded his medical institute; in New York, the Statue of Liberty was dedicated; in Tokyo, the Imperial University was completed. A sense of perpetual notion, of a life "chopped up by multifarious things" weighed upon
William James.

He and Alice had decided to buy a retreat in Chocorua, New Hampshire, "a bit of land on a lovely lake in New Hampshire, with a mountain
3500 feet behind it, and 90 acres of land, oaks, pines, etc., brook, water,
house." It had cost a hefty $750, but he thought the investment in con-
tentedness worth the price. James needed a getaway place; he was constantly stressed, headachy, and nervous. His book on psychology had *still*
to be completed; he was expanding his interest in philosophy; and he and
Alice were expecting another child. They wondered if they might not also
need a new, larger home in Cambridge.

The clutter of life and material necessity crowded into James's mind,
into his work. He felt so unwell that he'd resorted to consulting a "mind-
cure doctress." He had visited her eleven times for relaxation therapy: "I sit
down beside her and presently drop asleep, whilst she disentangles the
snarls out of my mind. She said she never saw a mind with so many, so
agitated, so restless, etc. She said my *eyes,* mentally speaking, kept revolv-
ing, like wheels in front of each other, & in front of my face, and it was 4 or
5 sittings ere she could get them *fixed.*"

Mind healers, faith healers, patent medicine salesman, all and more
were the bane of the American Medical Association, which had been working to eliminate all such "quackery" since its formation in 1847. The
AMA's first official action was to forbid its members to refer patients to lay
practitioners and anyone claiming to use homeopathic remedies. Its position was that only "scientific medicine," the stuff taught in universities and
practiced in hospitals, had any claim to legitimacy.

James was among many intellectuals, although few of them physicians, who found that view too restrictive. Another was Mark Twain, who characterized the medical establishment's efforts to eliminate alternative cures as pure economics: "The objection is, people are curing people without a license and you [doctors] are afraid it will bust up business," wrote the humorist. Twain contended that he could never get an honest answer from a licensed physician about alternatives, that it was "equivalent to going to Satan for information about Christianity." He was cynical enough himself to discount many claims as exaggerated, even dangerous. But Twain argued that patients should have access to all choices of treatment: "I want liberty to do as I choose with my physical body."

The insistence on scientific medicine, the ruthless effort to eliminate faith-based healing and other alternatives, paralleled the efforts of scientists to dust religious beliefs out of their methodology. James had never been easy with the idea that science should be so pure as to exclude all considerations of morality and philosophy. Neither did he embrace the "purification" of a single-approach practice of medicine.

Mind cures of the time featured a mix of relaxation techniques, hypnosis, and something then called "the talking cure," which would later be adopted by psychiatric followers of Freud and win a reinstated legitimacy as "psychotherapy." Mind healers might lack scientific rigor, James thought, but they were capable of insight: "The mind curers have made a great discovery—viz. that the health of soul and health of body hang together, and that if you get *right,* you get right all over by the same stroke."

He suspected that the mind-body connection was far more potent than nineteenth-century medical practice was willing to acknowledge. James had mixed feelings about his own particular mind-cure regimen, especially its failure to ease his insomnia. "What boots it to be made unconsciously better, yet all the while consciously to lie awake o'nights as I still do?" he wrote. Yet he opposed efforts to turn practitioners like his "doctress" into criminals. As Massachusetts moved to outlaw unlicensed medicine over the next decade, James infuriated his fellow medical school graduates by arguing eloquently—if unsuccessfully—on behalf of mind-body therapy. He followed by proposing that his own field, psychology, should study the power of mind over physical health and the "mystical stratum of human nature."

· · ·

JAMES GAVE *Phantasms of the Living* one of its first positive reviews, and the only endorsement to appear in a mainstream research journal. His account appeared in *Science* on January 7, 1887, and began, "This is a most extraordinary work." James went on to praise the intellect of the authors, their "untiring zeal in collecting facts," and their efforts to make sure that those facts were accurate. The book, he said, embodied "learning of the solidest sort."

Gurney's fear had been that the book would be ignored or underestimated, and when it appeared, he thought himself right. The book received only passing attention in the fall of 1886, mostly newspaper reviews by journalists who found the subject of apparitions to be an amusing topic. But by the next year, the research community had taken notice of *Phantasms'* claim to be legitimate science. Now the book received plenty of attention—excepting James's review and a polite notice in the philosophical journal *Mind,* all of it negative.

Perhaps the most publicized attack in England appeared in the magazine *Nineteenth Century* in August 1887, in a lengthy article devoted to discrediting the documentation that Gurney and his coauthors had used to establish their ghost stories. And perhaps the most painful challenge came from the United States in December 1887, in an article published by the ASPR and written by one of William James's close friends, Charles S. Peirce (the son of one of the Harvard professors sent to debunk the Davenport Brothers). Peirce had hardly bothered to read the book before declaring that he found every case of crisis apparition unbelievable, and Gurney's rebuttal showed that the critic had both misquoted and misrepresented the cases cited.

Still, Gurney admitted that Peirce had raised some valid points, notably that people are more likely to forget dreams or hallucinations that do not coincide with death, which could tend to inflate the statistics involved, and he repeated that he himself thought the statistical sample far too small. As he had tried to make clear, Gurney saw the book as only a beginning for psychical research—as did William James, who also noted that limited statistics and unanswered questions made it obvious that much more work needed to be done. James emphasized, though, that "any theory helps analysis of the facts." He encouraged Gurney to continue developing the theoretical relationship between telepathy and apparitions, despite its imperfections, and he pointedly stated in his review that in their combination of careful research and penetrating analysis, the authors had made a strong enough argument to finally draw the research community into the discussion.

"The next 25 years will then probably decide the question," James wrote. "Either a flood of confirmatory phenomena, caught in the act, will pour in, in consequence of their work; or it will *not* pour in—and then we shall legitimately enough explain the stories here preserved as mixtures of odd coincidence with fiction." He knew from conversations with Gurney that investigations of crisis apparitions had persuaded his friend he was studying scientific reality rather than science fiction. Despite the hostile response, James believed that as others explored the same territory, they too would recognize its genuine nature. He made that also clear in *Science:* "I feel that I ought to describe the total effect left at present by the book on my mind. It is a strong suspicion that its authors will prove to be on the winning side."

6

ALL YE WHO
ENTER HERE

TWO THINGS ONLY ELICIT AWE, said the philosopher Immanuel Kant, in 1788: "der bestirnte Himmel über mir und das moralische Gesetz in mir" (the starry sky above and the moral law within). To experience the former, one had only to gaze upward on a clear night. The moral law within required more effort. Kant suggested that morality could not survive unless people worked at it, stayed determined to maintain belief in three essential ideas: freedom, God, and immortality. There was a warning implicit in his declaration: one could rely always on a starry night; human morality was built of far more fragile material.

When Henry Sidgwick began studying philosophy, he found Kant's moral reasoning as powerful as if it had been said that day, as solid as the ground beneath his feet. As Sidgwick wrote in 1887, as a young philosopher he'd accepted completely Kant's doctrine, believed without question that "we must *postulate* the continued existence of the soul, in order to effect that harmony of Duty with Happiness which seemed to me indispensable to rational moral life."

Sidgwick's admiration for Kant and for the British philosopher John Stuart Mill, one of the most eloquent advocates of philosophy in service of society, illuminated his early writings. In his first book, *The Principles of Ethics*, published in 1874, Sidgwick explored the ways society might achieve the greatest happiness for the greatest number of its members. His next book, *The Principles of Political Economy*, published in 1883, was more pragmatic, exploring the ways philosophy might guide a society being reshaped by technology and commerce. There was less of Kant's higher morality in his work, more of Mill's utilitarianism.

Sidgwick found the years had eroded his ability to share Kant's acceptance of God as a premise, immortality as a postulate. By 1887, as he painfully acknowledged, Sidgwick needed some form of proof to support his faith. He knew that Kant himself had searched for evidence, as indicated by the German philosopher's investigation of the mystical claims of Emmanuel Swedenborg. But as far as Sidgwick could judge, even a century later, factual support for the God theory remained insubstantially thin.

Troubled by personal as well as philosophical doubts, Sidgwick found himself wishing for, at least, some sign of a "friendly universe." Perhaps this was what the psychical researchers could prove, that the star-spangled web of the skies was organized under some kind of greater moral governance, directed by "a Sovereign will that orders all things rightly." If the universe was only a machine, after all, what determined the rules of morality? What enforced them? Sidgwick found himself increasingly in agreement with Myers about the necessary nature of their search for proof of immortality. But he lacked his friend's natural buoyancy. He worried that they would fail. And he worried about the consequences if that happened.

"If I decide that this search is a failure, shall I finally and decisively make this postulate [to believe in God anyway]?" he worried to himself. "Can I consistently with my whole view of truth and the method of its attainment? And if I answer 'no' to each of these questions, have I any ethical system at all?"

IN EARLY 1887, as predicted by William James, the American Society for Psychical Research self-destructed, unable to withstand the hostility of its own scientific membership. James was keeping the remnants of the organization together, but he wrote unhappily to Edmund Gurney that as a Harvard instructor and author of a still unfinished psychology book, he really hadn't time to do so.

Gurney replied sympathetically. Of course, he understood that the "more legitimate, or at any rate, respectable work must oust the other." He had no doubt that James was keeping his priorities in the proper order. But the British society was not prepared to give up on its American counterpart. Sidgwick had received a cable from a wealthy American spiritualist. The man was a great admirer of Richard Hodgson's take-no-prisoners approach to investigation. The telegram indicated a willingness to pay Dick Hodgson's salary for a year, if he would come to America and put the psychical research program back on track.

Gurney didn't mention that Hodgson had refused when Sidgwick first told him of the offer. Or that Sidgwick, quietly relentless as ever, had persuaded his former student to go to Boston for the year only. He wrote only that James and the ASPR were about to find themselves lucky: "He [Hodgson] combines the powers of a first-rate detective with a perfect readiness to believe in astrology (*Don't quote this* as it might be misunderstood.) I should pity the astrologer whose horoscopes he took to tackling."

BUT ASTROLOGERS were not in Richard Hodgson's sights when he stepped briskly onto the Boston docks in May 1887. Leonora Piper was.

First, though, he had to restore some order to the ASPR shambles. Hodgson began by renting office space—two small, rapidly overflowing rooms on Boylston Place in Boston's bustling Back Bay—and hiring a part-time secretary. He planned to quickly reduce Mrs. Piper to the ranks of exposed impostors and, his American year thus well spent, return to England.

Although he had no way of seeing it, instead the move to the United States—and his investigations there—would be the most influential of his life, changing Hodgson, changing the essential nature of how he viewed reality. He was a man in a hurry when he arrived in Boston, with no sense of the supernatural mysteries—and the eerie possibilities—that would seduce him into remaining there.

In fact, Hodgson reminded James of a perpetual motion machine. The new ASPR secretary wrote letters; answered inquiries; met with callers;

attended committee meetings; researched "remarkable stories"; lectured on hypnotism to the Massachusetts Medical Society; attended unsuccessful thought-transference experiments by the surly, remaining ASPR scientists; and visited Ralph Waldo Emerson's grave to pay his personal respects.

"Well, I am happy enough," Hodgson wrote to his friend Jimmy Hackett. He liked the rocky New England shoreline, the curving green of the nearby Adirondacks. He liked Boston's sportsman culture, was already making friends at the Union Boat Club. He had his books of poetry and philosophy; he'd invested in a smart-talking green parrot for company. But he still dreamed of a wife and family.

He had moved around too much for that. "I should like to be married, if the fates permit, but have no one in my eye . . . if I am stirred profoundly by any woman, I expect I shall go for her like thunder, but I don't fancy I shall until then." In the meantime, he was concentrating on only one woman, an overrated medium with a reputation that needed exploding.

Leonora Piper shared with other mediums one particular convention of the time, a spirit guide. As was the custom, her "control" served as kind of spirit business manager, relaying messages, summoning other ghosts into conversations. Hodgson had never cared for spirit controls; from Madame Blavatsky's misty mahatmas to Florence Cook's astonishingly solid Katie King, they seemed to him implausible at best, fraudulent at worst.

Mrs. Piper's guide claimed to be a Frenchman named Dr. Phinuit, who had lived from 1790 to 1860. Shortly after Mrs. Piper went into a trance, her voice would change into his—deep, rough, flavored with a country French accent. Her personality would change too, from eager-to-please to abrasive, from gentle to forceful.

At least Phinuit didn't trail around in filmy garments flirting with visitors, but neither William James nor associates in France nor anyone else had found any records showing that the imperious Dr. Phinuit had ever existed. The control didn't even speak very good French. In conversations with James, who was fairly fluent, the doctor tended to rapidly fall into a baffled silence.

James suspected that the control was a creation of Mrs. Piper's subconscious, a fascinating mental process that seemed to serve to buffer her from the strangeness of the trance life.

Hodgson found Phinuit a silly complication and a "freak personality."

Still, with the personality's voice grumbling away during a trance, it was hard to ignore him. So Hodgson decided to simply confront the mess, interrupting to tell "Phinuit" that he was an obvious fake. The ever-irritable Phinuit promptly ended the sitting, announcing that he didn't want to talk to "this man" any more that day.

Hodgson rode the train back from the Pipers' suburban home to his downtown apartment, brushing cinders from his clothes and fuming about the whole deceptive, impossible profession of so-called mediums. But he returned, and so did Phinuit, this time prepared to put Hodgson in his place.

He had a message for Hodgson, he said, a personal one. The message came from a cousin, long dead, who was, according to Phinuit, the son of Hodgson's mother's brother. His grumbly voice continued: The cousin's name was Fred. He and Hodgson had gone to primary school together, played all kinds of rough-and-tumble games, of which leapfrog was Fred's favorite. As an adult, Fred had still loved gymnastic and athletic games: "He was swinging on a trapeze when he fell and injured his spine, finally dying in a convulsion. You were not present at his death."

Hodgson sat silent. But when he returned to his rooms, a small apartment he'd rented on Charles Street, he jotted down some notes. "My cousin Fred excelled any other person I knew at the game of leap-frog. He took very long flying jumps, and whenever he did so crowds of schoolmates gathered round to watch him. He injured his spine in a gymnasium in Melbourne, Australia, in 1871 and was carried to a hospital, where he lingered for a fortnight, with occasional spasmodic convulsions, in one of which he died. I was not present at either accident or death."

There was more. Phinuit told him of quarrels he'd had while traveling in Europe, of another lost friend, a slim, dark-haired woman—"she was much closer to you than any other person"—who had died in Australia, some years after Hodgson had moved to England. The woman wanted to be remembered to Hodgson's sister, who had been a close friend. And she had her own message for Hodgson. She wanted to make sure he kept a book of poems that he had given her and that her family had returned to him after her death.

Again Hodgson said nothing, although, in fact, the book sat on a shelf in his apartment, and he did plan to keep it. That night he contemplated a new set of problems. It seemed impossible to reconcile the extremes of

Mrs. Piper's trance—the Phinuit that she must have created, the information that seemed drawn from the air. The conclusion was obvious. She
must be spying on him—and most probably on her other visitors as well.
Hodgson decided to hire a private detective firm. He would have both
Mrs. Piper and her husband followed for the next month. Just in case, he
would tell no one of that decision, not even William James, and definitely
not Leonora Piper.

ALMOST FORTY YEARS had passed since the bright-faced little Fox sisters
first startled the country by demonstrating a new way of communicating
with the dead. In the more sophisticated 1880s—the decade of the electric
motor and the discovery of radio waves—Kate and Maggie Fox were long
past their heyday as darlings of the spiritualist movement.

Kate lived in New York City now, after many years in England, where
she had married a sympathetic barrister named H. D. Jencken and even
befriended Daniel Dunglas Home. Now widowed, she had a small apartment on the Upper East Side, where she conducted slate-writing séances.
In the 1850s, Maggie had become the common-law wife of a naval officer
and explorer named Elisha Kent Kane, famed for his expeditions into the
Arctic and his discovery of a channel that would eventually lead other
explorers to the North Pole. Kane had spent five years trying unsuccessfully
to wean her from spiritualism before his death in 1857. All these years after
his death, she still called herself Margaret Fox Kane. Like her sister, she had
worked as a medium throughout her life and still gave sittings in her Manhattan apartment.

Both women now struggled to hold an audience. Their spirit-rapping
technique had become an antiquated relic. Years of increasingly mediocre
demonstrations had cost them their following. Bitter, ignored, and impov-
erished, the sisters comforted themselves with alcohol. And both suffered
for it, in health and reputation. Acquaintances described Maggie as "a dissipated looking wreck." The few sitters who attended Kate's slate writing
evenings reported she was so drunk that she kept dropping the slates.

Nevertheless, a new scientific commission had asked to test Maggie's
powers, and she had agreed. Funded by a $60,000 bequest to the Univer-
sity of Pennsylvania from Philadelphia industrialist Henry Seybert, the

commission's stated purpose was to investigate the most credible supernatural claims and try to discover what lay behind them. Publicly, the Seybert Commission presented itself as a model of scientific objectivity. Privately, the scientists on the commission saw a good opportunity to further discredit the continued appeal of the supernatural. In a letter to James, the commission chair, University of Pennsylvania psychologist Howard Furness, described himself as a viper warmed by spiritualist nonsense.

With its first report, released in the spring of 1887, the Seybert Commission predisposition became apparent to all. The commissioners had chosen exactly the kind of mediums that the SPR thought untrustworthy: psychics for hire, professionals with a known act, from slate writing to ghostly materializations. They'd even made a point of investigating performers long discredited, such as the slate writer Henry Slade, now also known as a notorious alcoholic and a testament to the seedy nature of the professional medium trade.

As for the once-admired Maggie Fox Kane, the commission members found her an obvious joke. When she met with the commission members, faint thumps did shake the floor. They let the medium interpret them. She announced that the knocks came from the late Seybert himself, indicating his wish that the commission be thorough and patient. The members were unimpressed. In a more stringent test, they stood her on glass tumblers, two under each foot, effectively immobilizing her. This time the only person in the room who claimed to hear raps was Maggie Kane. In the commission report, her claim was followed by a stenographer's note: "No intimation is given that the rap here spoken of was heard by anyone other than the Medium herself."

In his concluding remarks, Furness sent his own message to other investigators, including James and his friends: Don't waste your time. "In my experience, Dante's motto must be inscribed over any investigation of Spiritualism and all hope must be abandoned by those who enter on it."

The report, published in the spring of 1887, produced a rare united reaction from the dedicated spiritualist community and the dedicated psychical researchers. Since serious investigations had started, the two groups had become increasingly alienated. True believers had resigned in mass from the British society after Hodgson completed his slate-writing exposé; Nora Sidgwick's response was that the group was better off without them.

American spiritualists had begun referring to "Professor James and his ilk" after his exposés of Boston mediums. James had responded by publicly accusing them of defending fakes "through thick and thin." But regarding the Seybert Commission, both sides agreed that the commission had only pretended to investigate, that the result was one more message of contempt from the mainstream science community.

James and his SPR colleagues worked, fruitlessly, to counter the bad publicity generated by the Seybert report. They wrote letters to their intellectual peers and to the press, pointing out that the Seybert Commission seemed to be deliberately investigating only those medium tricks already exposed time and time again. Why didn't the commission spend a little time and money on the phenomena that still proved puzzling? Myers, while politely praising the skeptical approach, suggested an investigation of automatic writing and other "perplexing phenomena which do admittedly occur but which need not be interpreted in the Spiritualist sense." And James wrote directly to Furness, urging an investigation of rare mental mediums such as Leonora Piper. Furness replied that he'd met with Mrs. Piper once and had not been impressed.

Anyway, Furness said, the terms of Seybert's bequest barred him from spending money on such experiments. He wouldn't bother to explain, he added in a sardonic postscript. Someday, when James had extra time, he could get Mrs. Piper's Phinuit to hunt up the dead industrialist and "get all the details directly."

THERE WAS NO LACK of grounds for such cynicism. As positive as the Sidgwick group was about their case for telepathy, they'd found plenty of fraud even in that area, thanks in part to the dogged and thorough Richard Hodgson.

Even from across the Atlantic, Hodgson had a gift for stripping away the claims of British psychics. Early in 1887, the Sidgwicks discovered a stage psychic who appeared to possess astonishing telepathic gifts. When Sidgwick wrote enthusiastically to Hodgson, though, his former student replied with a warning about some new tricks in the "telepathic communication" business. Thus armed, Henry and Nora discovered that the "acumen of Hodgson had already suspected the code of signals," which consisted mainly of a

confederate systematically sighing, groaning, or puffing out a certain number of breaths—three breaths for clubs, four for diamonds and so on.

The same vigilance had led the Sidgwicks and Gurney to make a
dismaying discovery about some of William Barrett's favorite telepathic
study subjects, three daughters of a Presbyterian minister. The girls had
been regarded as particularly upright in their family circumstances, and
some of Barrett's favorite experiments had involved them. These were the
girls who had so remarkably responded to his mental requests for oranges
and hairbrushes.

But in the ten years since Barrett first discovered them, Alice, Mary, and
Maud Creery's abilities had become far less remarkable. "The children
regretfully acknowledged that their capacity and confidence were deserting
them," Gurney wrote. Even worse, during card-reading tests, done shortly
after the publication of *Phantasms,* Alice and Mary were caught signaling
to each other.

Gurney had been worried about the girls' later test results. When carefully
watched, in situations similar to Barrett's first work in separated rooms, they
were now getting things consistently wrong. He decided to lay a trap, putting them in the same room, giving them an opportunity to cheat rather than
admit failure. During an evening at the Sidgwicks, Gurney and both of his
hosts observed the girls watching each other's eyes before "guessing" a card.

When confronted, Alice and Mary tearfully confessed an elaborate strategy of eye movements—upward look for hearts, down for diamonds, to
the right for spades, left for clubs, supplemented by different finger positions to indicate the value of the card. They confessed that if they couldn't
see each other, they signaled by noise: coughing, sneezing, loudly yawning,
and shuffling their feet. "I am very sorry to have to tell that we have
undoubtedly detected the two Creery girls . . . in the use of a code of signals to produce spurious 'thought-transference' phenomena," Sidgwick
wrote to Barrett. "Or rather two codes."

The SPR creed was strict on this point: once she was caught cheating,
every claim by a particular psychic or sensitive became suspect. Sidgwick
pointed out to Barrett that the girls' detailed code system obviously dated
back some time. He planned to strike all the Creery results from their catalogue of evidence. Barrett's pen fairly flew over the paper in angry defense.
In the first experiments, the girls did not act as agent and percipient; he

asked them to read *his* mind or that of another investigator. In those tests, the daughters had no forewarning and no one to communicate with. He doubted that they'd worked out signals for randomly chosen cups and saucers.

"I expect the natural alarm which the Cambridge Expts. has caused in our minds is probably apt to make us unjust in our judgment of the earlier experiments," he wrote to Sidgwick. "For my own part I am convinced that enough entirely trustworthy experiments exist with the Creery family to make it unwise to expunge the whole of their evidence."

The later, troubling results might be attributed, Barrett suggested, to what the psychical researchers would come to call the "decline effect," in which abilities seemed to lessen with time. One of the participants in Gurney's experiments on taste transference, the middle-aged owner of a Liverpool drapery firm, said that he could tell that his ability to "transmit" was weakening. He'd decided to retire from telepathy work. Almost all the subjects the investigators trusted reported the same thing, the transient, come-and-go nature of those mental connections.

The SPR investigators met to discuss this new challenge. Their subjects usually followed two different paths. The best ones quit after a few years, saying that it wasn't the same, complaining that whatever gifts they'd once had seemed to wear away. Others, reluctant to lose their clientele and reputation, developed cover strategies. One could even charitably make that case for the Fox sisters. But Sidgwick wasn't inclined to be charitable, not about the Fox family, and not in general. Fraud was the bane of every good result the SPR reported; it weakened every argument they put forth. A full retraction of all Creery experiments was ruthlessly published in the SPR journal.

The real problem, as Sidgwick noted in his diary, was not the lack of untarnished results. They still had plenty of those. He was more than familiar with "the transient glow of scientific enthusiasm." But the society lacked the sustaining warmth of a good explanation, a workable, testable theory for how that mental transfer occurred. That such a theory remained so elusive was troubling in the extreme. "If only I could form the least conception of the *modus transferendi!*"

IN BOSTON, the ever-organized Hodgson arrived at his ASPR office at 9:30 a.m. six days a week. The first mail always brought a dozen letters or

more. He promptly began reading and making notes. As instructed, at
10:30 a.m., Hodgson's clerk brought him a sheaf of typewritten letters dictated the previous day for him to read and sign. He then dictated answers
to the newly received letters. She mailed the signed letters. The two of
them dealt with correspondence and other office business until 1:00 p.m.,
when she took her remaining work to be completed at home.

He then dashed over to the Tavern Club next door for a quick snack,
usually hot tea and a dish of dates. He liked to schedule meetings and
interviews in the early afternoon, go back to the office, and scribble more
notes until the second mail delivery came. He'd take another quick dinner
break, but usually he hadn't finished making notes on those letters till late,
didn't get home till 11:00 p.m. or so, and even so brought a few more letters with him to finish. If he had time, he saved a little energy to write to
friends and read a little philosophy or poetry for pleasure.

He didn't complain to James or to the Sidgwick group back in England.
Only in private, to his old friend, did he acknowledge what his life was
becoming: "In one sense, I am sacrificing myself on the altar of psychical
research," he wrote to Jimmy Hackett.

At least he'd finally allowed himself one day off a week. He did not do
psychical research work on Sundays, as he once had, but tried to relax—
reading a biography of Shelley, going sledding with friends in the winter,
swimming and hiking in the summer, but often, after a week of psychical
research, he liked a solitary walk as much as anything. No wonder he had
no woman in his life.

"I am perfectly sick of seeing so many people and shouldn't visit more
than two or three if I were here as a private individual. But it is all for psychical research. Great must be my reward in heaven!"

The detectives following Leonora and William Piper had provided
Hodgson with a startling report. After a month of surveillance, they had
discovered nothing, absolutely nothing.

Neither Mrs. Piper nor her husband had been heard asking questions
about sitters; they'd had no mysterious meetings, made no unexpected
journeys, checked out no past issues of newspapers from the library, and
visited no cemeteries—all common practices of mediums gathering information about potential sitters. Further, Mrs. Piper had no detectives in her
own employ, busily supporting Phinuit's insights and explanations.

To Hodgson's surprise, Mrs. Piper did not find this clearance to be good news. She found it embarrassing and insulting. As she angrily told him, and then William James, respectable people did not find that detectives had trailed them around town. In a letter to James, she threatened to quit the whole research program. She was humiliated, and she was "sorely tempted" to have nothing to do again with his new ASPR secretary or any of the rest them. It took all James's diplomatic skills to smooth her down. He repeated his own respect for her, and teased her to see the silly side of detectives fruitlessly trailing such a sober couple, as William went to his department store job and Leonora ran errands to the baker and the greengrocer. "I hope neither you nor your husband will take the thing seriously. It has its very comic side and you are the ones who can best afford to laugh at it."

As for Hodgson, James told her, if they were ever able to understand her gifts, ever able to impress their critics with them, he was their best hope. James himself had been doubtful of the tough-minded Australian at first. But he had come round to Gurney's point of view. Hodgson, he wrote to Mrs. Piper, is "perhaps the most high-minded and truthful man I know."

LEONORA PIPER BECAME Richard Hodgson's personal obsession.

He paid the news seller in the Pipers' neighborhood to limit the family's access to information, only deliver morning newspapers on days when no sitting was scheduled. He hovered over the house like a bird on a nest, ignoring even the worst weather in order to maintain his watch. The winter of 1888 was one of the worst in history; the blizzards that swept the East Coast in March would kill more than 400 people. Hodgson remained undeterred. When conditions were so dismal that the Pipers' hilly street was just a glare of ice, he borrowed a sled and coasted back down toward the train station, three-fourths of a mile through the blurring cold.

He made a memorable and unnerving impression on five-year-old Alta Piper when he loudly lectured one visitor who had made the mistake of leaving a wet umbrella in the downstairs umbrella stand.

As she recalled it, he was shouting, "You idiot! Haven't you more sense than to do a thing like that? Don't you know you might be accused of being in collusion with Mrs. Piper if you leave your umbrella there?" It could be that she'd concealed a note in the folds, to be plucked by one of the daughters

and slipped up to the medium. And even if she hadn't done that, the possibility would ruin her sitting. "Bring your umbrella up, even if it is wet, and
in future mind what you're about."

Visitors couldn't use their real names or provide any personal information to Mrs. Piper. Hodgson sat in a corner, glaring, making sure they
didn't. Since the ASPR couldn't afford a stenographer, he took notes himself, pushing his bedtime back to laboriously transcribe his scribbles into
readable transcripts.

He was doing an admirable job, James said, but "one man can't do
everything, and is well nigh single-handed in the matter of investigation."
James knew he was providing little support. His publisher, Henry Holt,
was still berating him about the unfinished psychology book, and there
were the usual teaching and research duties at Harvard. He and Alice had a
fifth child, a baby daughter named Margaret. There was the apparently
never-ending job of remodeling the old house at Chocorua.

James deliberately made time where he could to meet with Hodgson
and even visit a few other mediums with the Australian. He was increasingly annoyed by the attitude of his own profession toward such work. The
Seybert Commission had squandered its opportunity, and the ASPR
researchers hadn't been much better.

"IT IS A MERCY that Hodgson exists," Gurney wrote to James in late May.
"I cannot help being glad that he is likely to stay a bit longer with you,
though he will be very welcome when he returns." Like James, Gurney was
mixing other interests with psychical research. He'd published a book of
essays on medicine and philosophy for the pure pleasure of writing out the
ideas. He had in mind next to do a book on hypnotism, which he thought
offered a fascinating way to understand the human mind as well as to
explore questions of mental communication.

He wrote of all those plans in an answer to a letter from James, who had
liked the philosophy book very much and praised its graceful prose and the
thoughtful way Gurney had linked questions of mind and health. The
message prompted Gurney to urge James to visit again soon: "Are you not
nearly due again on this side of the Atlantic? What a joy it will be to talk
again, such a number of new things to talk about! God bless you!"

It was barely a month later, on Saturday, June 23, 1888, that Gurney drove down to the resort town of Brighton, where he had been investigating a haunted house.

According to newspaper reports, he checked into the Royal Albion Hotel, a white wedding cake of a building directly opposite the bustling Brighton Pier with its carnival of penny mechanical games. He dined in the hotel coffee room and went to bed at about 10:00 p.m.

By two o'clock Sunday afternoon, he had not emerged from his room and had not responded to repeated knocking by both the maid and hotel manageress. They tried the door. It was locked from the inside. They called for the police to break the lock. The officers found Gurney dead in the rumpled bed, lying on his left side. His right hand held a small waterproof bag, a sponge bag used to hold toiletries, over his face. His mouth and nostrils were covered by it. A tiny bottle, containing a few drops of clear fluid, lay on the floor by the bed.

He was forty-one years old.

"ALAS! ALAS!"

On that Sunday, black and late on the day of discovery, Henry Sidgwick was sleepless with grief. He'd retreated into his study, lighting a single lamp, wrapping himself in the small pool of gold light. He sat by himself, writing in his diary, a private confessional of misery.

Fred Myers and his brother had brought "the terrible news" that afternoon. Arthur Myers, a physician, had gone to Brighton to identify the body, after police had found a letter addressed to him in Gurney's pocket. After informing Gurney's wife in London, he'd hurried to tell his brother, Fred, who'd collapsed into a numbed silence. The Myers brothers went together to tell the Sidgwicks. Fred Myers was still pale with shock and grief. He'd begged his weary brother to accompany him; he just didn't think he could carry the news by himself.

"I can write no more journal this month," Sidgwick wrote, alone in his quiet corner. "Fred Myers feels it terribly, but we too—Nora and I—do not know how we shall do without him."

James still had that warm note of invitation from Gurney in his office. He could hardly believe the news. "It seems one of Death's stupidest strokes," he wrote to his brother, Henry. As a colleague in psychical research, James thought Gurney, if anyone, could have carried the field forward, helped it to achieve respect. "I know of no one whose life-task was begun on a more far-reaching scale or from whom one expected with greater certainty richer fruit in the ripeness of time."

More, though, William James mourned a lost friend: "To me it will be a cruel loss, for he recognized me more than anyone, and in all my thoughts of returning to England, he was the Englishman from whom I awaited the most nourishing communion."

The Brighton inquest concluded that the death had been an unfortunate accident—an inadvertent mishap resulting from Gurney's regular use of chloroform to help him sleep. Sidgwick wanted to believe it.

Arthur Myers testified that he had known Gurney for nineteen years, both as a friend and as a physician. He'd treated Gurney for that reccurring neuralgic pain around his head and face. Gurney had been candid about his need to self-medicate. Dr. Myers had given him a number of prescriptions—morphine for sleep, sometimes chloral and belladonna for pain relief.

He knew that Gurney had also used the popular remedy chloroform to relieve neuralgia, traveling with it in case of an unexpected attack. Arthur Myers told the court that he was sure "Mr. Gurney had taken accidentally a larger dose than was his custom and had suffocated."

Sidgwick sought comfort in that certainty. "One more line," he wrote on June 29, returning compulsively to the secrecy of his diary. "I have just come back from the funeral. Arthur Myers, with whom I have had more than one talk, tells me that at the inquest on Monday there was a slight suggestion that it might be suicide; but it was easily and at once overborne by the evidence on the other side."

Alone in his study he admitted to "painful doubts." If his friend had been unhappy, wouldn't he have taken note? Wouldn't Nora have? "We saw him last on Tuesday 19th; he seemed to us well and in good spirits."

"THEY SAY THERE is little doubt that Mr. Edmund Gurney committed suicide," William James's sister, Alice, then living in London, noted in her own diary.

Friends were now talking about Gurney's mercurial nature, his ten-

dency to drive himself to exhaustion, his unpredictable highs and lows.
There were stories that he'd been driven to despair by psychical research,
his morale destroyed by all the criticisms of his work. That he'd been
demoralized by the amount of fraud they had encountered, such as the discovery of cheating by the Creery sisters; that he'd been depressed by the
response to *Phantasms of the Living.* The editor of the British philosophy
journal *Mind* told William James that he had worried that Gurney was taking his cause far too seriously, expressing concern about the "fury of this
hunt after ghosts and the like, which is positively wasting him, the very
body of him, I mean!"

There were slyer whispers, too, about his marriage. Kate Gurney had
spent many hours without her husband's company while the ghost hunt
possessed him. All her friends knew that she'd felt abandoned by Gurney;
now his friends began to wonder if he'd felt abandoned by his wife. As president of the SPR, Sidgwick sent Kate a formal sympathy letter in which his
sorrow yet leaked through—"nothing that can be said in public will really
express our sense of loss." He promised her that they would continue with
Gurney's work, not only because it was important; they "owed it to the
memory of our friend and colleague that the results of our previous labor
should not fail for faint-heartedness."

She answered their condolences politely; her warmest reply went to
William James, who she thought the most "akin" of Gurney's friends: "I
have a strong certainty that he is happier & still achieving. . . . I feel that if
I had never heard of the Immortality of the Soul—I should think *he* was
going on." But when the Society for Psychical Research established a
memorial in his honor, the Edmund Gurney Library, she declined to contribute. And as soon as the socially required year's mourning ended, she
remarried. The following year, Henry James saw Kate Gurney at a smart
little hotel in Paris. She was dressed in the latest style, the newer small bustle under her draped skirts, the little polished boots with their fancy buttons, the nicely tilted hat. She was no longer Kate Gurney, though; she'd
married a politician and publicist, Archibald Grove.

The couple was stopping in Paris for a few days on their way to a vacation in Tangiers. "How the drama of life rushes on," Henry wrote to
William. "And how out of it all poor chloroformed Edmund Gurney
seemed."

. . .

CHARLES RICHET'S HANDS flew like startled birds when he was excited, which was much of the time. Slight and intensely energetic, with a thin face, high cheekbones, and an enormous, wonderfully drooping mustache, the Paris physiologist was infectious in all his enthusiasms: the immune system, a complicated treatment he was trying against tuberculosis, an analysis of the mechanisms of fever—and, more recently, a growing interest in psychical research. He'd collaborated with the SPR on several telepathy experiments, impressing the members with his thorough methods and exuberant friendliness.

With Edmund Gurney gone, approachable scientists such as Richet had become more valuable than ever in the psychical research movement. Their numbers were perilously small to start, and if the movement was to survive, the interest and friendship of such people must be cultivated. Sidgwick moved to strengthen the connection with Richet, seeking information about his work with hypnosis.

Although a Scottish physician had coined the word *hypnosis,* the British tended to regard the practice as the stuff of superstition and quackery. So, in general, did the American scientists. By contrast, French researchers approached hypnosis as a science, working from a theory that it could achieve a specific neurophysiologic state, an unusual suspension of most mental activity. Gurney, who had been an early admirer of Richet and his work, had considered the research gap as he mulled over a possible book on hypnosis. As he'd told James ruefully, "I don't know how it is to be done. The French have all the material!"

In Paris, well-known physiologists Jean Charcot and Pierre Janet (the latter an associate of Richet) were using hypnosis to treat nervous system disorders, including epilepsy. Charcot believed that hypnosis tended to tranquilize the nervous system, creating what appeared to be a dreamlike trance. The French scientists could not agree on how to define the trance itself. Charcot and some of his colleagues, working from a strict physician's point of view, felt that it indicated an abnormal mind, so that similar effects should not appear in the mentally healthy. Psychologists argued that, done well, hypnosis could induce a trance in anyone, that a hypnotic trance revealed the ability of one mind to impose its will on another. They pointed out that people with no evidence of mental defect *had been* and *could be* hypnotized. It was in studying the healthy, psychologists said, that the most interesting questions could be asked, such as why some minds appeared more suggestible than others.

As it pertained to psychical research, Richet considered another aspect of hypnosis. He wondered whether the trance state relaxed mental barriers, thus allowing one mind to become more open, more responsive to the thoughts of another. It occurred to him that a genuine medium might create her own internal hypnotic trance, that perhaps Mrs. Piper's Phinuit was a personality produced by self-hypnosis.

Certainly, there was evidence linking the hypnotic state to telepathy, some of it from Gurney and Myers. Richet had also conducted earlier experiments, hypnotizing patients at a Paris mental hospital. He still remembered one adolescent girl who'd been sent there for treatment. One day Richet had brought an American medical student with him on a hospital tour. After Richet had hypnotized the girl, he said to her, "Do you know my friend's name?"

Of course she didn't. She began to laugh.

Let's make it easier, he said; "Look, what is the first letter of his name?"

She lay in the narrow cot, silent, apparently asleep. He was ready to walk away when she replied, softly, "There are five letters; the first is H, then E . . . I do not see the third, the fourth is R, and the fifth is N."

The student's last name was Hearn.

Richet calculated that the odds of the patient getting it right by chance were something like two hundred thousand to one.

Only recently, his colleague Pierre Janet had hypnotized a patient named "Leonie" and sent her "traveling" away from her secluded country home. Janet had been testing the idea that in a hypnotic trance, the human mind could float, balloonlike, drifting to the location that its owner wished to visit. Leonie announced that she was going to Paris to visit M. Richet.

Suddenly, her voice sharpened and rose. She said, "It is burning!" She began to twitch and toss in distress. Janet moved to calm her down. Leonie relaxed into a sleeplike state again, but unexpectedly, anxiously opened her eyes, saying, "But M. Janet, I assure you that it is burning." When Janet contacted Richet, he learned that the Paris laboratory had caught fire that morning and burned to the ground.

If there wasn't a connection between the hypnotic trance and the medium trance, then Richet didn't know his science. He agreed too with one of Gurney's ideas, that hypnosis gave the scientist control over his subjects, reducing the possibility of cheating. The son of a doctor, trained in medicine at the University of Paris, Richet was confident that science and its techniques could easily manage and even manipulate the supernatural.

It would take a temperamental Italian peasant woman with an apparent gift for summoning the wind to make Richet wonder if that assumption was, perhaps, a little arrogant.

ON AUGUST 9, 1888, an obscure Italian physician from Naples posted an open letter in a leading newspaper of Rome. He addressed it to his country's most famous psychologist, Cesare Lombroso.

Lombroso, known to his countrymen as the Master of Turin, had built his reputation on the study of criminal behavior; he was one of the first psychologists to propose that killers were born, made by biology, an argument that would gain even greater appeal when the science of genetics came of age.

Lombroso's 1876 book *L'uomo deliquente* (Criminal man) explained that natural killers were easily recognizable: stupid, small-skulled, heavy-browed, throwbacks to the club-bearing early humans. His notion of phrenology—the science of reading a person's behavior in the shape and size and even the bumps of his skull—would rapidly become one of the most influential ideas in Victorian psychology.

Lombroso was notoriously hostile to spiritualism; he'd publicly made a point of siding with both John Tyndall and T. H. Huxley in their indictments of occult beliefs. But he was also known as a thoughtful man; he might believe that criminals were born, not made, but he also advocated humane treatment of prisoners. He was passionately against the routine use of capital punishment.

The doctor in Naples hoped merely to catch his famous countryman's interest. "I want to say something about a patient," wrote Ercole Chiaja in the *Fanfulla della Domenica,* "a sick woman belonging to the lower ranks of society and who is now about thirty years old." The woman in question had been orphaned at thirteen and was wild almost beyond control. She refused to learn to read or write. She refused to take daily baths. She liked to drink; frequented bars on the docks; picked up sailors, a different one each night if possible. Her friends tried to cultivate her mind "with unremitting patience but without avail."

Eusapia Palladino had the face and body and waddling walk of a bulldog. She was uneducated and flamboyantly promiscuous. She made furniture fly. She caused marks to appear on paper by merely extending her hand. Tied to a chair, she caused fingerprints to appear in a smooth block of clay across a room. Those were only a few of the bizarre occurrences that swirled around her, the doctor said.

Chiaja realized that this sounded improbable—no, impossible. But he had *seen* these things, all of them. He had only one request: he wanted Italy's best psychologist, the great Lombroso, to tell him whether he, Chiaja, was crazy or sane.

MANY PEOPLE WOULD be glad to leave the year 1888 behind. It had blown in on the ill winds of a lethal winter, lapsed into a tragic summer, and it would leave on a note of homicidal insanity. From August until November, a faceless killer nicknamed Jack the Ripper haunted the streets of London, butchering five prostitutes before disappearing like a demon in the night.

Against such murderous drama, one might think that the continuing downfall of the Fox sisters would barely register. But for better or for worse, they were still the stuff of newspaper sales. In October 1888, desperate for money, Maggie Fox Kane sold a confession to the New York World, a declaration that the Fox sisters had built their careers on a rare ability to loudly pop their toe joints. The resulting rattle first fooled their parents, she said, then their neighbors, then the rest of the world.

The month after the article appeared, Maggie Fox—washed, sober, neatly dressed—rated a stage demonstration at the New York Academy of Music. As her sister Kate, who was also there, wrote a friend, the hall overflowed with hostile faces, ill wishers happy to see the Fox sisters fall from grace. The event's managers cleared a good $1,500. Maggie saw little of that, and Kate saw none.

The *World* reporter planned to turn the confession into a book. He was going to call it *Death Blow to Spiritualism,* and he expected it to sell very well. Kate was overwhelmed by outraged spiritualists demanding that she prove her sister wrong.

She thought she might do that—see if there was a little money in it, find out if there was any chance of salvaging their reputations. "They are hard at work to expose the whole thing, if they can; but they certainly cannot," Kate said, although she acknowledged that many former friends thought that she and Maggie were now traitors to the cause.

BY THE SPRING OF 1889, with the wretched past year behind them, Sidgwick and Myers were determined to return the momentum to psychical research. They'd organized a volunteer workforce of more than four hundred SPR members to carry out Gurney's plan for an expanded Census of Hallucinations. They were determined, as Sidgwick put it, that Gurney's "six years labour should not be lost." *Phantasms of the Living*—following James's prediction—had set off a new swell of interest in their work, a membership stirred by a vision of possibilities beyond the bumbling of the professional medium circuit. The first order of business was to do that larger statistical sample and a more rigorously controlled survey.

They had a straightforward question they planned to ask: "Have you ever, when believing yourself to be completely awake, had a vivid impression of seeing or being touched by a living being or inanimate object, or of hearing a voice; which impression, so far as you could discover, was not due to any external physical cause?"

Every SPR volunteer was asked to put the census question to at least twenty-five adults. No one from Gurney's earlier survey could be included. Anyone who claimed a history of seeing visions was automatically eliminated. Volunteers were encouraged to query among groups likely to contain a disparate assortment of people, such as workers at a factory, residents of an apartment complex or office building, guests at a dinner party. They hoped to present their preliminary findings at the International Congress of Experimental Psychology, meeting in Paris that summer. Psychical research wasn't normally part of the program, but thanks to Richet's influence, the organizers had included a session for the SPR's very experimental branch of science.

It was a heady moment, and it reinforced a sense that, despite obstacles and discouragements, they were making progress. Psychical researchers from all over Europe, from Asia, from North and South America, planned to attend. Sidgwick had been invited to speak. So had Myers. And so had William James.

James planned to give some account of what Richard Hodgson was doing with Leonora Piper. Although Hodgson's year of donated salary was over, Sidgwick and Myers had taken over paying him. Obviously, American scientists had not rallied to the work, but as Myers said, to give it up now would be "deplorably hasty."

Hodgson was building a detailed picture of work with a credible medium, leaving out nothing, documenting every disastrous sitting, every dubious encounter, and every moment of dumbfounding accuracy. His reports were terse, to the point, and made fascinating reading:

Miss Mary A. T. Sitting: This was a complete failure.

Mr. X. Sitting: This was a complete failure.

Mrs. H. O. Sittings: "My first sitting was not satisfactory, there seemed to be much guessing." Further, as Mrs. H. O. left the house afterwards, she met some friends going into the house. On her return, the medium gave full names of her relatives with accuracy "but nothing was given that these friends did not know."

Mr. A. Y. Sittings: "At the first interview several remarkable phenomena occurred. Although I was introduced by another name, my true name was early given and some incidents of my life stated which could not have been known to the medium."

Mr. E. D. C. Sitting: "The communications I had through Mrs. Piper were of such a nature, I should hardly like to put on paper. I will say, however, that I went there totally unknown to her, and the names she called and the facts she spoke of, known only to myself and those who are no longer here, astonished me beyond measure, for I had never before visited a medium or seen anything of the kind."

Dr. C. W. F. Sittings: "At my third sitting, 'Phinuit' said: 'William Pabodie sends his love, says he has suffered remorse of conscience and if he had to live his life over again, he would not do what he did.'"

The doctor had driven from Rhode Island for the sitting. William Pabodie was a friend of his who had committed suicide almost twenty years earlier.

Mr. J. R. R. Sitting: He handed the medium the collar of a dog he had once owned, asking "Phinuit" if there were dogs where he was. "Thousands of them," he said, then suddenly he cried, "Here is your dog coming, a long way off. You call him.

"I gave my usual whistle.

"Here he comes. . . . Rover, Rover, no G-rover, Grover, that's the name."

The dog was once called Rover, but was changed to Grover in honor of the 1884 election of President Grover Cleveland.

And then there was the sitting with William's wife, Alice, and his younger brother, Robertson James, on the morning of March 6, 1889. The brothers' aunt Kate—their mother's sister—had been ill, and Alice asked about her health.

"She is poorly," Phinuit replied shortly. Alice confessed to being disappointed by the baldness of the answer. She was just planning to ask again when the medium suddenly threw back her head and said, in a startled way, "Why Aunt Kate's here. All around me I hear voices saying, 'Aunt Kate has come.'"

When pressed, Phinuit told them that Kate Walsh had died very early that morning. He just wasn't sure about the precise time, maybe around 2:00 a.m. or so.

The sitting made Bob James so nervous that he went directly from Mrs. Piper's house to Hodgson's office in downtown Boston. As he'd hoped, he found his brother William there, deep in conversation. On hearing the account, Hodgson immediately wrote a statement of what had happened, noting date and time, and adding, "This is written before any dispatch has been received informing of the death." He insisted that it be signed by all three of them.

The meeting broke up shortly, and James returned home. A few hours later, he received a telegram from his cousin. His aunt Kate had died early that morning, a few minutes after midnight.

It was one more mystery for William James to take to the meeting in Paris.

"I AM QUITE thick now with Sidgwick, whom I like amazingly, odd as he is," James wrote to his wife from the Congress of Experimental Psychology.

Sidgwick had charmed James with his unself-conscious sweetness. James had enjoyed watching the fifty-year-old British philosopher keep up his exercise program, jogging down the Paris streets with coattails flapping behind him. He'd admired Sidgwick's shy effort to overcome his stutter: "It was pitiful to hear him try this a.m. to express himself in French, with his bad choice of words, his bad accent and his fearful stammer. Myers did quite well, and I not so badly."

All three of them, and Richet as well, argued to expand the Census of Hallucinations into an international program, resulting, James said, in a "somewhat stormy" session. The "orthodox" scientists resisted the idea strenuously, even ones that they'd expected some support from, such as Pierre Janet, Richet's colleague in hypnotism studies. Janet explained that he didn't want to see his research associated with ghouls and ghosts and goblins. In the end, James and his friends mostly prevailed. The Swiss, the Germans, the Italians, and—unexpectedly—the Brazilians had agreed to conduct surveys in their countries as well. James was now, somewhat to his dismay, the official coordinator for the U.S. part of the census. A report was due at the next meeting of the congress, in 1892.

As James wrote to his wife, he'd found the session much more energizing than the usual research conference: "The whole thing is amusing and exciting to the last degree—so much that I have got quite out of the state of torpor into that of jiggle and am lying awake at a bad moment." At least the insomnia gave him a moment in which to write home.

. . .

"A CURIOUS CHAPTER might be written on the pseudo-confessions of mediums," Charles Richet wrote late in his life, ruminating on the downward trajectory of the Fox sisters.

As a class, mediums were notoriously unstable, he noted. Their abilities, if genuine, were unreliable and erratic. Even the good ones tended to eventually suffer from the decline effect. The weirdness of their profession often produced mental problems, if those didn't already exist. The whole realm of psychic powers impressed Richet as transient, an ephemeral grasp on something in the air, which came and went, one of the problems with studying them. The Fox sisters, late in life, seemed to encapsulate every quality that made mediums so difficult.

In November 1889, Maggie Fox Kane confessed again, retracting her earlier confession to the *New York World.* This time, she insisted to the *New York Press* that her statements about faking the spirit sounds had been "false in every particular." She had made up the story for money, she said. She had done it for "promises of wealth and happiness in return for an attack on Spiritualism," and she was sorry to say that she'd reaped neither of those rewards. She wanted readers to know that she had not been paid to grant this later interview; she hoped merely to undo the damage she had done to her own reputation, to her sisters, and to the credibility of her profession: "Would to God that I could undo the injustice I did the cause of Spiritualism."

Richet suspected that the truth about the Fox sisters lay somewhere in that welter of claims and denials. It was possible that when young and relatively unspoiled, the Fox girls had possessed some mediumistic talents. It was possible that whatever small ability they possessed had failed long ago, lost in the circuslike promotion that surrounded them.

It would have demanded extraordinary courage, not to mention upright character, for these mediums in particular to confess to a decline effect, to announce that their much-heralded abilities had vanished long ago. Richet didn't find Kate and Maggie particularly upright; he doubted they had ever been notably talented. They were, however, a cautionary tale to mediums and investigators alike: "That the Fox sisters, after the enormous developments of spiritualism that followed on their early demonstrations, should have tricked is possible or probable, not to say certain."

. . .

IT WAS SOME weeks after the death of James's aunt that Mrs. Piper suddenly grabbed Hodgson's right hand. Her grip was painfully tight; Phinuit seemed to have suddenly disappeared. The voice speaking was that of a woman, terrified, speaking in a desperate whisper.

"Help me, help me," she said. "I'm so cold." She tugged at him with her right hand, pulled until he grasped her left hand.

"That hand's dead—dead—this one's alive." Her left hand seemed icy cold to him; her right had been hot and clammy.

"Who are you?" he asked

"Kate Walsh," she replied

It was so different from the other Piper sittings, Hodgson told James. It was intimate and, somehow, terrifying. "It was the most strikingly personal thing I have seen." James followed up by writing to his cousin Elizabeth, daughter of his late aunt Kate Walsh, and asking if any of it made sense to her.

Yes and no, Elizabeth wrote back. Her mother had suffered a partial paralysis as she lay dying. One of her hands had been "dead," but Mrs. Piper had the side wrong. It had been her right side that was paralyzed, not her left. It might be that the medium was describing a spirit still feeling the effects of a last illness, but the sensations had come through somewhat garbled. "Queer business," James wrote to Hodgson, eerie and uncomfortable in its effect.

And yet, as they both knew, mediums were unreliable creatures. The Fox sisters had reminded them of that, and in spades. Perhaps he and Hodgson had become too close to Mrs. Piper, too seduced by the rather sedate contrast she provided to the generally seedy community of mediums. Perhaps they were losing their perspective. Perhaps despite their precautions, she had found a way to outwit them, here in Boston on her home turf.

Almost simultaneously, James and Hodgson came up with the same idea. They would send her to England, put Gurney's "intractable Atlantic" between her and any regular sources of information and support. They would see what happened to their "Yankee girl," as James affectionately called her, in a place not her home.

7

⌐yⱤ⌐

THE PRINCIPLES
OF PSYCHOLOGY

T. H. HUXLEY—among the most intellectually confident of nineteenth-century men—had no difficulty, was proud even, to confess that he felt ignorant in one notable area. He first made his confession at an 1869 meeting of the Metaphysical Society, a club of Britain's leading intellectuals, with membership dominated by theologians, scientists, and philosophers, spanning John Tyndall to Henry Sidgwick.

The Metaphysical Society held discussions ranging from "The Scientific Basis of Morals" to "Has a Frog a Soul?". Huxley's talk was titled "Is God Unknowable?", and in it he described his perspective on religion as "agnostic" (from the Greek *agnostos,* "unknowable"). Where the gnostic writers of Christian history claimed spiritual knowledge as essential to salvation, this champion of scientific materialism claimed no knowledge. There could be no knowledge, Huxley believed, when there was no proof.

The more nineteenth-century science illuminated mysteries of the universe, the less likely it appeared to Huxley that solutions to those mysteries lay beyond physical reality. Besides that, his labors on behalf of Darwinian science had left him less than impressed with the intellectual powers of his fellows. Was it even possible that this meager species, saddled with finite perception, finite intelligence, could comprehend a reality unbounded by physical limits? He doubted it. Huxley insisted that the rational mind must investigate, must seek hard evidence. But in the absence of such evidence, he viewed doubt, not belief, as the responsible stance.

Twenty years later, in 1889, Huxley took that confession of unbelief even further, in a widely read book, *Christianity and Agnosticism,* which made it clear that the intervening time had reinforced his doubt, intensified his impatience with those who refused to join him in doubting.

Agnosticism, Huxley wrote, should not be considered a creed. It was a method, or if one preferred, a scientific principle: nothing is certain unless it is proved. Thus, if one could not confirm the literal existence of angels, spirits, gods, they must be doubted. The agnostic did not close his mind to possibility; he simply set the standards of reality high. Very high. No supernatural agency had met Huxley's burden of proof.

Indeed, Huxley saw science as directly contradicting supernatural explanations. Astronomers had thoroughly overturned the concept of an Earth-centered universe, once essential to Christian thought. Throughout the nineteenth century, Earth's importance—and correspondingly, mankind's place in the universe—had further diminished. This planet was one of eight in our solar system (the most distant, Neptune, having been discovered in 1846). Our sun was one of millions, if not billions, of stars in the Milky Way galaxy, which now appeared to be itself one of countless galaxies. Rather than the center, God's favorite, one of a kind, Earth appeared but an insignificant part of a vast material universe. Each glimmering star, each celestial object found, Huxley wrote, revealed another "world in course of development or retrogression hanging out its light signal in the ocean of infinity."

Physical man, Huxley continued, was "mere dust in the cosmic machinery, a bubble on the surface of the ocean of things both in magnitude and duration[,] a by-product of cosmic chemistry. . . . He fits more or less well into this machinery, or it would crush him. But the machinery has no more special reference to him than to other living beings." In the debate over whether the scientific worldview should replace religion, Huxley considered the answer already given. It had.

William James, Henry Sidgwick, and their fellows, although they also counted themselves as rationalists, could not go nearly so far. To exclude from reality anything not demonstrated through the scientific method was to accept on faith, they would argue, that there *is* no reality beyond what a select group of people (on an insignificant planet) say is so. To deny the existence of the spirit—without thoroughly exhausting the subject through dogged research—to accept such arbitrary limits, was to them a prejudicial view, closed-minded and *un*scientific.

Yet as much as the psychical investigators—led by philosophers, after all—were motivated by grand principles, by the metaphysical generalities at the center of their quest, they constantly found their energies drained by small particulars. In the fall of 1889, the particular problem was the stubborn behavior of a certain Boston medium.

Mrs. Piper rejected all calls to duty or to higher purpose and flatly refused to travel to England. Yes, she could see the importance to advancing psychical research. She appreciated the researchers' enthusiasm for her abilities. As she kept repeating, however, her husband, William, "would not hear of being left alone, subject only to the ministrations of a housekeeper." Her daughters, Alta and Minerva, were too young to be left without a mother. She was sorry, but family took precedence over explorations of spirit communication.

To change her mind required weeks of effort, the combined persuasive talents of James and Hodgson, their exhortations that Mrs. Piper consider the contribution she could make to history, to human knowledge, and to the understanding of her own abilities. Even so, she said a reluctant yes only after the SPR agreed to send her children along with her on the journey. Her husband's parents helped the cause by inviting him to stay at their home in the new Boston suburb of Arlington Heights during his family's absence.

Fred Myers wrote a letter that sang with relief: "Dear Mrs. Piper—I am so very glad you are coming after all! And we will do our best to make your visit pleasant." Seeking to boost her spirits, James invited her and Hodgson for a pre-departure visit to his country home in the New Hampshire countryside she loved so well. Chocorua was outstandingly lovely that October. The surrounding wooded hills were set with the gemstone colors of autumn— amber and carnelian, garnet and citrine—heart-stoppingly brilliant against a clear blue sky. They spent the days actively, out of doors, saving serious talk for the evenings. Mrs. Piper hiked with James and fished with Hodgson.

(Since James disliked family fishing expeditions, the Australian was in the habit of taking James's children out on Lake Chocorua to try for bass.)

The quiet, sun-touched days confirmed both men's impression of their subject as unguarded, basically uncomplicated, nice. Nothing about her pointed to a schemer, a confidence artist. Her powers, whatever they were, seemed as much a mystery to her as they were to James and Hodgson. The medium side of her life made her uncomfortable, fearful of appearing foolish, uneasy about what happened during her trances. "I should be willing to stake as much money on Mrs. Piper's honesty as that of anyone I know," James wrote to Myers in a burst of sudden optimism following that golden week in New Hampshire.

MRS. PIPER AND her daughters sailed on the Cunard liner *Scythia*, leaving Boston on November 9, 1889. The day was bright and clear, the wooden docks noisy with the clatter of wheels, stamping of hooves, and reverberating calls of "Godspeed." At the water's edge, women pulled their gloved hands out of velvety muffs to wave good-bye; men waved their hats of beaver felt and stiffened silk. Against the glitter of sunlight on water, the three Piper females picked out William ruefully smiling his good-byes from the dock below.

They pressed against the rail as if to cling to him, to home. Alta sobbed. Her father threw some new bright pennies up to her. Years later, she would remember trying to catch them, the copper coins slipping through her fingers. She saw her father's smiling face through a teary blur. Minerva tugged at her hand as the boat slid away from the dock, as the slice of blue water widened. She didn't look away from him, but she could hear Minerva beside her: "Don't ky, Alta, please don't ky; you're making mama ky too."

The *Scythia* carried them from an unseasonably balmy New England autumn to a frigid North Atlantic winter, where the liner had to mince its way between drifts of ice. When the trio arrived in Liverpool, Mrs. Piper was sick with a miserable cold. They checked into a small hotel, and all three tumbled into a warm, shared bed.

Two days later, Fred and Evie Myers came to meet them. Myers had sent a description of himself: "I am rather tall and stout, with short, grayish beard and probably great coat with fur collar." Evie was dressed in her favorite dark blue, a matching hat tilted elegantly over her pretty face.

Mrs. Piper met them in her gray dress with the lace collar, her ash-brown hair caught sedately up behind her head. Her daughters hovered near, brushing her skirts with their white pinafores, big-eyed and shy. What Alta remembered years later was the relief embodied in the meeting: "Why Mrs. Piper," exclaimed Evie Myers. "You are not at all like what I expected. I thought you would wear your hair in frizettes and be dressed in magenta."

Fred Myers's reaction was less obvious, masked by the friendliness of his welcome. But he too felt a gust of relief blow through him. As he would write later in an autobiographical essay, the sober little American medium stirred in him a sensation like the rising warmth of a hot-air balloon. He defined the feeling as hope.

THE PROSPECT OF finding a real psychic at last also caught the attention of Oliver Lodge, a physicist considered one of the SPR's more promising researchers. Lodge was a classically trained scientist with a natural bent toward rebellion.

The son of a prosperous pottery merchant, he had refused his chance at the family business. Born in 1851, young Oliver had spent countless nights during primary school staying up too late, mapping the stars. In adolescence he had turned the family kitchen into a chemistry laboratory. As a young man, he'd built a basement workshop where he constructed a voltage regulator out of his father's pocket compass, a battery from the pieces of a blown-down weathercock.

In the early 1870s, Lodge had moved to London to study physics at University College and attend lectures at the Royal Institution. For the rest of his life, he would remember, and be troubled by, the sensation of participating in a religious orthodoxy there. The Royal Institution was regarded as a sacred place, where "pure science was enthroned to be worshiped for its own sake. Tyndall was in a manner the officiating priest and Faraday a sort of deity behind the scenes."

Lodge found himself increasingly uncomfortable. He respected completely the precise laws of physics, the elegant formulas of chemistry. In science there was undeniable truth, great power. But he did not see in it the Truth or an omnipotent Power deserving of worship. As much as he valued science, it was not to him the only system of thought, not a catch-all replacement for all previous belief systems. Unlike so many of his colleagues, Lodge did not view science and spiritual belief as mutually exclusive, incompatible ways to understand reality. Ever the rebel, as his father could have told his colleagues, Lodge rather wished to mend those divisions, even dared to wonder if science could be used to help explain a spiritual reality beyond the material.

His sense of unease, mixed with curiosity, had led Lodge into his early telepathy experiments and into the Society for Psychical Research. The deeper he got, the more intrigued he became.

Rebellion aside, by the late 1880s Lodge had acquired the requisite imperious manner of an established researcher, a manner to go with his station in life. There was something imposing in the intensity of his dark eyes, in the alert angle at which he held his bald head with its fringe of prematurely gray hair. Chairman of the physics department at University College, Liverpool, Lodge was also happily married and the father of twelve children (a fact at which William James always marveled). Progressive in his social attitudes, the inveterate tinkerer Lodge encouraged his daughters as well as his sons to experiment in their father's home laboratory, unfazed by both minor explosions and—he reported proudly—one large one that blew out windows in neighboring homes.

At his university laboratory, meanwhile, Lodge was pursuing one of the most exciting new ideas in physics. In 1873, the great Scottish physicist James Clerk Maxwell had put forward a theory that electromagnetic waves could be produced in a laboratory. In 1887 German physicist Heinrich Hertz successfully demonstrated Maxwell's theory by both generating and receiving such invisible waves. Lodge was at the forefront of physicists envisioning these "Hertzian waves" harnessed, put to use in sending and receiving messages. By the end of 1889, he had built a crude, short-distance wireless telegraphy system based on Hertz's work. He would soon design an improved device for detecting Morse code signals transmitted wirelessly. Called the coherer, Lodge's invention would become a standard part of the earliest radio receivers, including those used by the young Italian Guglielmo Marconi.

Despite such pursuits, or maybe because of them, Lodge remained intrigued by psychical research. He thought of his coherer as an exceptionally acute ear, able to receive invisible waves that might carry messages. Could this concept of unseen paths of communication—wavelengths yet undiscovered—explain some of the most stubborn questions in psychical research?

In telepathy, Lodge speculated that one person's mind might be sending such waves, another's receiving them. Perhaps a first-class medium such as Leonora Piper was an exceptional natural receiver, a better, finer-tuned listener than the average person—a "coherer" in the psychical sense.

When Myers began to organize Mrs. Piper's visit, Lodge thus saw an opportunity. He promptly offered to host her visit, exasperating his wife Mary, who, as she confessed later, feared she would be entertaining some kind of carnival freak. Lodge dismissed such concerns. It was a chance to test his theories of mental transmission firsthand, against the best mental medium so far known.

He proceeded to further try Mary's patience with fanatical preparations for the medium's arrival. He insisted on replacing the entire staff of their home, hiring new servants before the Piper family arrived. He locked away the family Bibles and photograph albums. Upon their arrival, he searched the Pipers' luggage. He insisted on reading all their mail before they saw it. He monitored the most casual conversation at meals, driving Mary crazy with his insistence that she talk about nothing but general current events— Gustave Eiffel's tower, erected at the World Exposition in Paris, or the record winter tormenting Europe. He dedicated his study as a place for the sittings and put a paper sign on the door reading "SILENCE." Years later, Alta Piper still remembered with amusement Lodge's two-year-old daughter cautioning, as they trooped by, "Hush! Papa's smilence is on the door."

Alta remembered playing with the Lodge children and cozy evenings of stories and hot milk served with cake before bed. For her mother, however comfortably housed, it was a very different kind of stay. Despite Myers's promises of a "pleasant visit," being a medium under scientific investigation guaranteed long hours and physical discomfort. No medium would say different. The Fox sisters had been searched to the waist, tied at the ankles, balanced on pieces of glass; Anna Eva Fay had been bound to metal rings; Florence Cook's hair had once been nailed to the floor so that

William Crookes could be sure of her immobility. Leonora Piper suffered as much and worse at the hands of her supposedly gracious hosts.

Lodge—and Fred Myers, who had come to help with experiments—began by testing her trance state. How deep was it? Could it be penetrated, broken by sensations? The men pricked her with pins, burned her arm with a match, held ammonia under her nose. Nothing seemed to disturb the sleeplike daze. They also learned quickly, as James and Hodgson had before them, that a credible trance didn't guarantee good results from the improbable Dr. Phinuit.

Some days Phinuit had nothing to say. His rumbling voice did no more than fish for answers, ask questions of the sitters, repeat what they had told him as if it were an incredible discovery. Anyone who sat with Mrs. Piper only once, and arrived on one of Phinuit's bad days, was likely to leave disenchanted, Lodge said. But as the investigators discovered, the transcendental eeriness of the good days could make one forget that.

Lodge persuaded a physician friend to visit. The doctor informed him, resentfully, that he would come for friendship but would do nothing helpful, not say one word to this so-called medium. He would give the same response to anything she said—correct or incorrect. He planned to do nothing but grunt.

Phinuit told the doctor that he had four children, one a little girl, aged thirteen, a "little daisy" who had beautiful dark eyes, with a curious little mark or scar over one eye, and who unfortunately was lame. The man also had a boy who was not so sweet and who should be sent to school for his own good. The doctor himself drank hot water when he had indigestion. Oh, and Phinuit said the doctor had recently had a bad experience, when he "nearly slipped once out on the water."

The doctor stalked out in silence. It was only later that he told Lodge that almost every fact was correct. He and his wife had been discussing whether the boy should go to school. The doctor had been in a dangerous yachting accident that summer. His daughter, Daisy, was dark-eyed, with a small scar over her left eye from a tumble as a baby. But she was not lame.

The doctor returned for a second visit. Lodge introduced him, simply, as the man who was here yesterday.

Oh, Phinuit said, he'd made a mistake the day before. The man's daughter had a friend who was lame. Daisy was deaf.

And that was exactly right. His little Daisy had lost her hearing after a fever. The doctor decided not to come back. He didn't want to know what Mrs. Piper might reveal next.

Still, Lodge pointed out to Myers, the doctor's surprising encounters with the medium, and other similar successes, didn't prove spirit communication. The psychic researchers could be merely gathering evidence of an exceptional telepath at work. Perhaps Mrs. Piper just picked up stray thoughts from the physician or others standing near. Lodge thought they needed to go further, to create a situation that would challenge her to provide information unknown to anyone in the room.

He had an elderly uncle, Robert, living in London. They weren't particularly close. But since they were family, Lodge would ask a favor of his uncle. Robert's twin brother had died twenty years earlier. Lodge would write to his uncle and ask for an object belonging to the dead brother. The Americans claimed Mrs. Piper had a gift for psychometry. He and Myers would see what she made of whatever object turned up.

"By morning post on a certain day I received a curious old gold watch, which this brother had worn and been fond of; and that same morning, no one in the house having seen it or knowing anything about it, I handed it to Mrs. Piper when in a state of trance."

She turned the ornate gold watch over in her hands, over and over yet again. This belongs to one of your uncles, she told Lodge in her rumbling Phinuit voice. The owner of the watch was very fond of another uncle. The name of the other was Robert. In fact, Robert was now the keeper of the watch.

Her hands moved the gold timepiece back and forth, restlessly. Her voice changed; when she spoke again, it was smoother, quieter. "This is my watch and Robert is my brother and I am here. Uncle Jerry, my watch."

The dead uncle's name was, indeed, Jerry. Over her head, Lodge exchanged an excited glance with Myers. Still, this might yet be no more than another exceptional telepathic demonstration. Lodge knew the name; it was readable in his thoughts.

Tell me something then, said Lodge, something that only you and your brother Robert might remember. Some story that I couldn't possibly know.

She continued to turn the watch in her hands, carved back to carved front, over and over, in a ceaseless golden circle of motion.

Phinuit began to talk of the brothers swimming a creek together when they were children. It was dangerous. They came close to drowning once. They once killed a cat in Smith's field. Jerry remembered that as a child he owned a small rifle. That he'd treasured a long, peculiar skin. Phinuit thought it was a snakeskin.

Lodge immediately wrote again to his Uncle Robert, asking if any of these stories were correct. His uncle wrote back that he definitely remembered their risky swims in the creek. And he still had his brother's prized snakeskin. He wasn't too sure about the rest of it, but then, his memory wasn't what it used to be. Perhaps his younger brother would recall more.

So Lodge wrote next to his other uncle, who replied with evident surprise. He remembered all of those events: the way the creek flowed past a treacherous millrace, the stiff action of the rifle, and even the poor cat, trapped in Smith's field. He wanted his nephew Oliver to know that he was not particularly proud of that story, and that the brothers had tried to keep it a secret.

But secrets leak. If his uncle remembered, Lodge reasoned, so might others. Perhaps, despite all their precautions, Mrs. Piper had managed to track down these odd details of his uncles' long-ago childhoods, knowing that he might inquire about family members. Lodge sent a private investigator to the town where his three uncles had grown up, to find out whether she could have culled the information somehow from sources there.

"Mrs. Piper has certainly beaten me," the investigator wrote back. There was no indication that anyone else had made such inquiries. And even if they had, nothing in the local records or newspaper back files contained the details that she had provided. Apparently, she had just snatched them out of the air.

IN THE SAME autumn that saw the Pipers traversing the icy Atlantic, Cesare Lombroso, the Master of Turin, gave in and decided to visit Naples, after all. He would expose this so-called medium, who refused to disappear from public view, whose exploits continued to appear in newspapers, to be talked over at spiritualist meetings.

In late November 1889 Lombroso checked in to a luxury hotel and demanded that Eusapia Palladino be brought to him. When the sturdy little woman entered, swathed in black, he regarded her unsympathetically. He wanted her tied to a chair, he said flatly. He'd brought thick strips of linen with him for that purpose. He himself would turn up the lights; he wanted the shadows banished to the bare corners.

Trussed like a turkey, the medium sat, muttering a little to herself. Lombroso watched her, stiff with distaste. Impatient after some ten minutes of nothing, he rose to get rid of her. But then the curtains lining a window some three feet away began to move, as if a wind was rising beneath them. They blew toward him, away from the window, which was set into the curve of an alcove. A small table, tucked into the alcove, began to slide toward the medium. Two of Lombroso's colleagues immediately dove behind the heavy velvet curtains to grab her confederate.

No one stood there.

The windows were locked tight shut on the inside, just as Lombroso had secured them before the woman had arrived. The curtains continued to blow in their invisible wind, and the little table kept chugging along toward her.

Lombroso stomped over to find the attached wires. He found none. He thumped for hidden passages, checked for hidden machinery. Still nothing. Just Eusapia Palladino, staring mockingly back at him, her wrists still bound to the carved wooden arms of the chair.

"I am filled with confusion," Lombroso wrote. Unwilling to concede, he decided, like the Neapolitan doctor before him, to seek help.

.   .   .

"IT SEEMS TO ME," William James once wrote to a colleague, "that Psychology is like Physics before Galileo's time—not a single elementary law yet caught a glimpse of."

That lack of knowledge had prompted his book—and enormously delayed it. But, finally, after twelve years of work, on September 25, 1890, James's publisher, Henry Holt, released a fat, two-volume set titled *The Principles of Psychology.* (He would later release a condensed, one-volume version, leading psychology students to call the long version the "James" and the short version the "Jimmy.")

In *The Principles of Psychology,* James had set himself the daunting task of supplying those missing elementary laws, trying to give shape to this still amorphous science, a discipline that had arisen as traditional religion and philosophy no longer seemed adequate to explain human behavior. A thoroughgoing scholar, literate in four languages (English, French, German, and Italian), he drew on a wide range of other, longer-established disciplines and ideas. First, he applied his training as a biologist and a physician. The book laid out causes and effects of behavior as delineated by modern science, based on a developing understanding of the nervous system as it processed reasoning, memory, association, and emotion. He enriched it with lessons from philosophy—the idea of self, the concept of truth, the necessity of moral behavior.

Just as its author's free-ranging mind reveled in the messy complexity and contradictions of human behavior, so did *The Principles of Psychology.* In writing it, James also allowed himself to be guided by his own observations of the relationships between mind and body, between individuals, between self and society.

"No one," he wrote, "ever had a simple sensation by itself. Consciousness, from our natal day, is of a teeming multiplicity of objects and relations, and what we call simple sensations are results of discriminative attention, pushed often to a very high degree."

He was the first psychologist to propose a feedback mechanism between emotions and physical sensations related to them. "Objects of rage, love, fear, etc.," he wrote, "not only prompt a man to outward deeds, but provoke characteristic alterations in his attitude and visage, and affect his breathing, circulation, and other organic functions in specific ways." He also believed that emotion could be as easily triggered by memory or imagination as by the event itself, a concept that would decades later form the foundation of post-traumatic stress disorder. Or in James's words, "One may get angrier in thinking over one's insult than at the moment of receiving it."

Our individual reactions, he said, could never be separated from the broader fabric of human life. The intensity of response to James's hypothetical insult, whether that response were delayed or immediate, would depend on the relationship in question. One person created countless different relationships in this life, each based on the perception of him by another and by his response to that perception, so that "a man has as many social selves as there are individuals who recognize him and carry an image of him in their mind." Going a risky step further, James also suggested the possibility of another series of relationships, those that operated outside the observable limits of material reality.

Although he never intended *Principles of Psychology* as a reply to Huxley's agnosticism, James's book did present an alternative vision. It offered a world less mechanical, more infused with human possibility—including the possibility of psychic phenomena.

In this landmark psychology text, James discussed trance personalities, telepathy, spirit possession, even Leonora Piper. He didn't, as Alfred Russel Wallace had done, declare psychic phenomena to be proven laws of nature. But he did emphasize that if one wished to understand the human mind, it was necessary also to understand why such phenomena were seen and experienced by so many people.

Accessible and provocative, the book elevated James to a new level of public stature. Copies were purchased not just by psychologists and their students but also by a wide spectrum of readers fascinated by this new science of behavior. The jurist Oliver Wendell Holmes Jr., then on the Massachusetts Supreme Court, wrote, "Dear Bill: I have read your book—every word of it—with delight and admiration." A fellow philosopher wrote, "I am not overmuch given to hero-worship but [with this book] you certainly have been my hero."

The reception by the science community included some less enthusiastic appraisals. Stanley Hall, head of psychology at Clark University and one of the more disgruntled former members of the ASPR, wrote to the author to praise "your magnificent book." But to others, Hall complained that there was a little too much William James in it for his taste. And the pioneer of the European experimental psychology movement, William Wundt, was coolly dismissive: "It is literature. It is beautiful. But it is not psychology."

Fred Myers had nothing but praise for James's "big and good book." More than a publication, however, he saw *The Principles of Psychology* as an opportunity. In an exuberant letter, Myers wrote to acquaint James with that opportunity, to make sure he realized that there was no other SPR member "on the whole so well situated as you for the successful pushing of the inquiry."

"I believe that with a view to (a) the good of mankind (b) even to your ultimate fame, it is essential that a main part of your energy shall henceforth be devoted to these SPR inquiries," he added.

Even across the wind-scoured miles of the Atlantic, James wrote back, he could recognize in Myers's letter the voice of a "despot for psychical research." Graciously demurring, he countered that Myers himself was far more useful to the cause. His British colleague possessed a gift for ensnaring others in his enthusiasms. "Verily you are of the stuff of which world changers are made."

Although not prepared to devote himself to the subject, James agreed that his position—"Professorship, book published & all"—provided "a good pedestal for carrying out psychical research effectively." As he assured Myers, he had no intention of wasting the platform. James had that year also published in a popular magazine an essay favorably comparing psychical research to other fields of science. To the readers of *Scribner's,* James had extravagantly praised his colleagues, naming Henry Sidgwick "the most incorrigibly and exasperatingly critical and skeptical mind in England." He also praised the SPR publications: "Were I asked to point to a scientific journal where hard-headedness and never-sleeping suspicion of sources of error might be seen in their full bloom, I should have to fall back on the *Proceedings of the Society for Psychical Research.*" The group's work he described as cautious, meticulous, and wonderfully puzzling. In that assessment he included the British investigations of hallucinations and hypnosis, as well as Richard Hodgson's detailed analysis of Leonora Piper.

The medium and her daughters had with great relief returned to Boston earlier in the year. After taking some months to reestablish her household, she had only recently permitted séances to resume. Hodgson, James wrote, "is distinguished by a balance of mind almost as rare in its way as Sidgwick's," capable of accepting that supernatural events could be verified and equally capable of shredding pretense of such occurrences. "It is impossible to say in advance whether it will give him more satisfaction to confirm or to smash a given case by its examination."

James admitted that the psychical researchers had yet to prove the existence of supernatural beings or powers. He thought, however, that they'd made a credible start toward that goal. He also thought that reading the SPR journals would lead any reader to conclude that mainstream scientists were wrong to dismiss and denigrate such work. In perhaps his strongest criticism of those scientists to date, James wrote in *Scribner's* that in its determined orthodoxy, scientists had come to seem a mirror image of those clergymen who insisted on only one way of seeing the world: "Science means, first of all, a certain dispassionate method. To suppose that it means a certain set of results that one should pin one's faith upon and hug forever is sadly to mistake its genius, and degrades the scientific body to the status of a cult."

EARLY IN THE SUMMER of 1891, William James's sister, Alice, wrote from her London home to warn her brother that doctors had discovered a tumor in her breast. He wrote back immediately, with his own unique idea of encouragement and advice.

Like all the James children, Alice had long been prone to ill health, the same "neuralgia and headache and weariness and palpitations and disgust" that were often William's companions. Perhaps more than any of them, she suffered from miserable bouts with depression, what William called the "nervous weakness, which has chained you down for all these years."

He wrote, "I should think you should be reconciled to the prospect [of death] with all its pluses and minuses. I know you've never cared for life and to me now at the age of nearly fifty, life and death seem singularly close together in all of us." He assured Alice that for all her weaknesses, she also had amazing strengths—"fortitude, good spirits, unsentimentality"—in the midst of grief and illness. His work with Mrs. Piper, with "enlargements of the self in trance," as he put it, encouraged him to believe that some people might achieve their best potential after death. He hoped his sister would find comfort in that: "When that which is *you* passes out of the body, I am sure there will be an explosion of liberated force and life until then eclipsed and kept down." James imagined his sister's entrance into the other world as something dramatic, a shock of energy and light.

"It may seem odd for me to talk to you in this cool way about your end; but my dear little sister, if one has things present to one's mind, and I know they are present enough to *your* mind, why not bring them out?"

In late July, James received a letter from his sister's companion, Katherine Loring, telling him that the cancer was malignant and spreading. Alice was sedated with morphine; she'd asked Loring to reassure her brother that the pain was not terrible. She also wanted him to know that whatever the future held, she was not unhappy with herself or her life: "when I am gone, pray don't think of me simply as a creature that might have been something else."

In late September, spurred to do more than share musings on life and death, James arrived in London. Alice teased that it was her "mortuary attractions" that had coaxed him across the ocean. Gradually, though, her first burst of cheerfulness faltered. "She talks death incessantly," he wrote to his wife. "It seems to fill her with positive glee."

He stayed two weeks before returning home. Alice was nauseated by the morphine. He recommended hypnotism to counter that, patiently teaching Loring how to induce a hypnotic state. But his sister fought the very idea. She considered hypnosis part of William's psychical nonsense; it reeked of their father's mystical ways and clutching spirits. She hadn't found his metaphysical promises reassuring in the least. "I hope," she confided to her diary, "the dreadful Mrs. Piper won't be let loose upon my defenseless soul."

IN THE DECEMBER 1891 issue of *Harper's,* Mark Twain published a personal endorsement of the science of the supernatural. Twain began by declaring that the Society for Psychical Research had accomplished what many said could not be done. It had made the study of the occult a respectable endeavor.

Further, Twain said, the SPR pioneers had freed people like himself to speak out on such subjects, in this case, on telepathy (which he liked to call mental telegraphy). He had tried to write about it earlier, he said, figuring that his own reputation would be enough to sell such a piece. But his editor had flatly refused to glorify the occasional "coincidence" as telepathy. Now, he wrote, the hard workers of the SPR had "succeeded in doing, by their great credit and influence, what I could never have done—they have convinced the world that mental telegraphy is not a jest, but a fact, and that it is a thing not rare but exceedingly common. They have done our age a service—and a very great service, I think."

Twain's intention was to offer his own evidence to support their work, in particular his personal experiences with telepathy. He told of a visit to Washington, D.C., which involved a very late arrival. He knew that a good friend was also planning to be in the capital; but "I did not propose to hunt for him at midnight, especially since I did not know where he was stopping."

Although it was late, Twain found himself restless. He went out for a walk, drifted into a cigar shop, and stayed for a while, "listening to some bummers discussing national politics." Suddenly his friend came back into his mind, with startling specificity. If he left the shop, turned left, and walked ten feet, his friend would be standing there. Twain immediately walked out the door and turned left. There was his friend, standing on the edge of a street corner, chatting with another man, delighted to see Twain stepping up to join the conversation.

"In itself the thing was nothing," Twain commented. "But to know it would happen so *beforehand,* wasn't that really curious?"

The essay went on to catalogue other such events: of many times thinking of one friend or another and writing a hasty note, only to find that the friend had written to him at nearly the same time. Of hearing his wife suddenly mention an event that had just crossed his own mind. Of two writers almost simultaneously coming up with the same idea for a story; of two inventors creating a similar device in almost the same month. He proposed that telepathy could even account for scientists such as Darwin and Wallace developing their insights into evolution during a similar time period.

"We are always mentioning people, and in that very instant they appear before us. We laugh, and say, 'Speak of the devil' and so forth and there we drop it, considering it an 'accident.' It is a cheap and convenient way of disposing of a grave and puzzling mystery. The fact is it does seem to happen too often to be an accident."

Twain proposed that most people pick up the occasional thought from someone else, casually, telepathically, without conscious awareness. They simply underestimate or deny their own ability. "While I am writing this, doubtless someone on the other side of the globe is writing it too. The question is, am I inspiring him or is he inspiring me?"

The following month, an inspired response appeared in *Scribner's*, an article titled "The Logic of Mental Telegraphy," by which the author, Joseph Jastrow, clearly meant the illogic of Mark Twain.

Jastrow, of the University of Wisconsin, had been one of the first members of the ASPR and one of the earliest dropouts. Most of the former ASPR members now preferred to ignore their old association. But a few, notably Jastrow and his mentor, Stanley Hall of Clark University, wished instead to dismantle it.

Hall, founder of the *American Journal of Psychology*, wanted to cut psychology clean of any link to the theological study of behavior. Jastrow, a former Hall student who'd founded the psychology department at Wisconsin, was a pure researcher to the core, a noted experimentalist in the field of visual perception. He'd earlier been responsible for those ASPR experiments that discredited the claim that mediums possess a rare sensitivity to magnetic fields.

Both Jastrow and Hall felt that mere withdrawal from the psychical research society wasn't enough, that they were obliged to atone for their earlier sins by exposing its wrongness. They worried that adopting their peers' more lofty approach—ignore it, and it will go away—meant waiting too long to weed superstition out of science.

As for the demonstrable existence of telepathy, wrote Jastrow furiously in *Scribner's*, "nothing could be farther from the truth." If Mark Twain perceived that he lived in a world too full of coincidences, he could be excused. He was only a writer, a former journalist at that.

In the late nineteenth century, Jastrow explained, with so many people connected by telegraph and telephone, traveling on fast trains and steamboats, people crossed paths as never before. So did their thoughts and ideas. It was hardly surprising that some thoughts occurred simultaneously; it was a natural response of people receiving information at nearly the same time. Jastrow described Twain as a typical spiritualist, insisting on the supernatural explanation when an ordinary one would do: "He detects mysterious laws of fortune and freaks of luck . . . and utterly refuses to believe the general doctrine of chances, because it is not obviously applicable to his particular case."

There was one little coincidence involved in writing his essay. Jastrow had read Twain's article while on a cycling tour of New Hampshire. Stopping at a library to consult a road atlas, he had seen the latest issue of *Harper's* and been so outraged that he went right to his hotel and wrote to the competition, the editor of *Scribner's,* proposing a counter article. As soon as Jastrow completed the trip and collected his mail, he discovered that the editor had simultaneously sent him a letter, asking for a response to Twain's telegraphy article.

Jastrow was a meticulously honest man. He told that story himself, included it in the article. But he cautioned the reader not to make anything of it. His mind had not communicated with that of the editor. It made perfect sense that they might both happen upon the same notion when seeing Twain's article. It could happen thrice over, and he would think no different.

Sidgwick might be hardheaded, Hodgson obsessive, Jastrow hostile in approaching the occult. None of them was armed, however, with all the cold, bare facts reported in a new insider exposé of spiritualism, published in 1890 by an author known only as "A. Medium."

Thoroughly impressed by it, Hodgson would spend years trying to find the anonymous author of *Revelations of a Spirit Medium,* but to no avail. A. Medium never surfaced to face the hundreds of spiritualists enraged by this perceived betrayal. *Revelations* was a manual—albeit a very funny one—on how to gull the gullible.

Take, for instance, the eerie lights that sometimes graced séances. A. Medium had a recipe for that: Take an empty two-ounce cough syrup bottle and fill it one-fourth full with water. Cut the heads off about one hundred parlor matches, drop them in the water, and cork the bottle. Once the phosphorus had dissolved off the match heads, remove the floating bits of leftover pine from the resulting brownish muck and recork the bottle. In a dark room, when the cork was removed and a little air let into the bottle, it would become "a beautiful yellowish luminous shape."

"Try it, reader," wrote the author. "You will be astonished at the results you can obtain from a bottle of this 'cough mixture,' a white handkerchief [draped over the glowing bottle] and a dark room."

If a reader wanted to able to handle hot coals—as D. D. Home had sometimes done—A. Medium had a formula for that too: one-half ounce of camphor, two ounces of aquavitae (filtered water), one ounce of quicksilver (mercury), and one ounce of liquid styrax (a natural solution of myrrh), shaken well, spread over the hands, and allowed to dry. You could then "hold your fingers in the blaze quite a while without any bad effect."

The writer offered tips for materializing spirits. There were shoemakers who would fit shoes with hollow steel heels for only $20. A. Medium advised filling one heel with fine white netting, to be draped over the body, and the other with an assortment of cloth masks "with which to transform your own face a dozen times."

The book recommended a stash of faces painted on cardboard, which could "peer" through the curtains of a cabinet. It told of, but did not endorse, the practice of one female medium who had painted a baby's face on one of her breasts and pushed it out between cabinet curtains to be kissed.

A. Medium found the joint-cracking explanation of the Fox sister's rapping phenomena ridiculous "in view of the fact that there are much more simple methods." You could sit at a table with your hands resting on top and your thumbnails just touching. "Press them together tightly and slip them a little a time. You will find every time you slip them, one against the other, quite a loud 'rap' will be heard." Books, slates, knees, and heels could all produce rapping sounds as well.

The book presented tips for almost every possible result produced by a physical medium. And that was its point. A. Medium professed weariness with fraud, with the "thousands of persons earning a dishonest living through the practice of various deceptions in the name of spiritualism." The author still considered himself a believer. He expressed a certainty of life beyond the grave, of seeing friends in an afterlife, and "more than likely" a physical return to Earth. But for this, A. Medium would not require the help of professionals who were cheats, frauds, drunken men, women "no better than a common prostitute."

In the end, A. Medium made the same recommendation that the SPR had been making for years: Investigate the spirit world, but avoid paid mediums. Remember that any street conjurer possesses the tricks to make lights dance in the dark, tables walk in the air.

THE AMBITIOUS CENSUS of Hallucinations was coming together, but slowly. The SPR's best statistician, Nora Sidgwick, now had another demanding job, principal of Newnham College.

Both Nora and her husband had worked to see this new, all-women college established at Cambridge. So had Nora's family, including brothers Gerald and Arthur Balfour, who had donated money to see Newnham built. The college's chemistry laboratory was named after another Balfour brother, Francis, who had died in a mountain-climbing accident. As soon as Nora was offered the appointment, Henry Sidgwick knew that his wife would accept and throw herself wholeheartedly into the job.

It would have been uncharacteristic of her, however, to abandon the unfinished Census of Hallucinations just because she had new responsibilities. Nora left for her office early each day, dressed in her favorite simple black, her hair pulled back, exuding calm. She looked much the same—still neat and composed but undeniably tired, her husband thought—as she came home each evening, only to return to the census.

"If you ever find me getting slack about the SPR, you must pull me up," she told her assistant.

Sidgwick considered his wife the brightest of his circle. He was delighted at her latest achievement, but worried that she couldn't possibly—and shouldn't—keep up the double pace of college administrator and psychical researcher. Yet how would the organization manage without her? They'd already lost Gurney, whom William James once candidly described as "the *worker* of the society," the one best able to get through "drudgery of the most colossal kind." Sidgwick did not think the group could afford to let Nora go.

"I fear she may not find time for the work of the SPR, for which I think her uniquely fit—much more fit than I am," he wrote. "If it turns out that she must sacrifice some of this work, I shall have to take her place; but my intellect will be an inferior substitute."

He proved better suited to helping with the social chores of college administration. He served as a reliable companion at official dinners, receptions, and even meetings with students. The ever-serious Nora still had not mastered the art of conversation. Despite practicing on Henry, she found herself reduced to monosyllables during party chitchat. Worse yet, jokes tended to turn her silent. Henry stepped in to fill the gaps. "He could talk nonsense," she said with genuine admiration, "or subjects."

The once awkward Sidgwick had surprised himself and his friends by growing into rather a charming man. Despite his shyness and his stammer—still with him, if less pronounced—he'd learned how to disarm others with his self-deprecating discourse and diffident humor. His gentle wit had even gained a nickname: Sidgwickedness. He was a good listener, a man who obviously considered the opinions of others. Sidgwick became the SPR's chief diplomat, working to smooth its way to the next International Congress of Experimental Psychology. His formidable task was to dispel resentments still lingering from the previous session, at which some scientists had felt pressured into allowing the touchy subject of psychical inquiry. Sidgwick paid courtesy visits to the influential university psychology departments in Berlin and Paris, assuring scientists there that at the upcoming congress, he planned to organize a small "orthodox" session on hypnotism only. He also tried to make his case that the SPR's subject matter was not an aberration and that telepathy, as "a law of nature," should be considered a legitimate subject for science.

The meetings with French and German scientists didn't exactly encourage him. He encountered rigid resistance from what he called "stubborn materialists, interested solely in psychophysical experiments on the senses." They weren't in the least impressed by Henry Sidgwick, a philosopher whose whole experience in experimental techniques was *with* telepathy.

"Water and fire, oil and vinegar, are feeble to express our antagonism," he said gloomily to Nora.

IN THE FIRST WEEK of a gray and blustery March, a dying Alice James asked her brother Henry to cable her good-byes to her family in America: "Tenderest love to all farewell. Am going soon." The next day, Henry cabled again: "Alice just passed away painless."

To Henry's annoyance, William promptly wired back to ask if their sister was really dead. She could, after all, have slipped into a trance state: "her neurotic temperament & chronically reduced vitality are just the field for trance-tricks to play themselves upon."

Henry had been every day to the sickroom to comfort his beloved sister. "My patient Henry," she called him. He was not in a mood to put up with his older brother's transoceanic speculations. "If you were here," the novelist snapped back, "you wouldn't have thought your warning necessary."

"What a relief!" William replied, sobered. Their sister was at peace, the "task" of her life completed. For this once, for the sake of Alice and for a peeved and grieving Henry, the eldest James sibling dropped the subjects of trances and spirits in favor of a simple good-bye.

"GOD (OR THE UNKNOWABLE) bless you," James wrote to Hodgson, referring to Huxley's agnostic ideals.

As the summer of 1892 approached, the James family prepared to sail for Europe. Harvard had granted William another sabbatical leave, and he intended to use it for a good year of travel, study, and visiting friends. This time his wife and children would come with him to the Congress of Experimental Psychology. But they would also visit his brother in England and go to France, Switzerland, Germany, and any other place that might appeal to them.

Before he left, James wanted to make sure that the American part of the Census of Hallucinations was ready. The tireless Nora had sent him a "skeleton paper" so that he could see her methodology and an outline that he could use, showing how to organize the American cases along similar lines. He left her correspondence in Hodgson's room along with his "hallucination book," containing a fair number of analyzed crisis apparitions, and two boxes of unanalyzed cases. He feared it would be up to Hodgson to finish the analysis and was sorry to leave so much work.

Hodgson was also polishing the final draft of his Piper report. James hoped that, using Nora's guide, the ASPR secretary could largely "fill in the blanks" for Mrs. Sidgwick and turn the whole pile of ghostly accounts around in a month or so. But he knew that the reports, like all human experiences, were complicated and undeniably messy. "Heaven help you anyhow," James offered ruefully. "You'll be troubled with duplicates and ambiguities enough."

If only there were more Hodgsons, thought Sidgwick. It wasn't just that Nora was so busy. In addition to her Newnham administrative chores, she was also finishing a study countering claims that higher education was unhealthy for women. He, meanwhile, continued developing his utilitarian philosophy; James was traveling on sabbatical; Podmore remained a busy postal inspector; and Myers, a school inspector. Lodge worked away on the physics of wireless communication. Charles Richet, who had always devoted the bulk of his time to physiology research, had now taken up a hobby, building a glider in which he could pinwheel through the air around a family chateau.

"No one is saying—as Hodgson in America—'Psychical Research is the most important thing in the world; my life's success and failure shall be bound up in it,'" Sidgwick mused in his diary. "Yet I am convinced that only in this temper should we achieve what we ought to achieve." It was only through the day after weary day of Hodgson's kind of work that the unexpected suddenly revealed itself.

In the midst of a sitting with a young married couple, Mrs. Piper, lost in a trance, talking in her deep Phinuit voice, had demanded that the watching Hodgson leave the room. This needed to be a private conversation.

"I want to talk to you about your Uncle C.," Phinuit said, unusually gently for him, according to notes taken by the woman.

"Is he in the body?" the woman asked, although she knew perfectly well that her uncle was dead.

"No," said Phinuit.

"How did he die?"

"There was something the matter with his heart and with his head. He says it was an accident. He wants you to tell his sister. . . . He begs you for God's sake to tell them it was an accident—that it was his head, and that he was hurt there (makes motion of stabbing heart), that he inherited it from his father. His father was out of his mind—crazy."

The control, the medium, one of them was somehow describing the woman's family in Germany. As the sitter wrote to the ASPR, her grandfather had been mentally disturbed after falling from a horse and injuring his head. He'd remained half-crazy for a good three years, until his death in 1863. His oldest son, her uncle, had committed suicide, thirteen years after his father's death, in "a fit of melancholia" by stabbing himself in the heart.

The family was ashamed of his insanity and the unpardonable suicide. Her family never spoke of it. Only two people in America, besides her husband, even knew the story, and those two people lived nowhere close. "It is absolutely impossible that Mrs. Piper got at the facts through information derived from these persons."

Hodgson published that account, and accounts of many of the earlier sittings, in his first report on Mrs. Piper in the June issue of the *Proceedings of the Society for Psychical Research.* He knew that there was an audience waiting to hear what the famed debunker of psychics would say, an audience wondering if he would expose her as he had Helena Blavatsky, or if he would declare that he now believed in talking to the dead.

Hodgson managed, neatly, to disappoint expectations.

He admitted Mrs. Piper's trance to be genuine. He'd put it to every test that he could reasonably conduct. He'd put ammonia-soaked cloth under her nose, dumped spoonfuls of salt, perfume, and laundry detergent into her mouth, pinched her until she bruised, all without provoking a flinch. Mrs. Piper sometimes complained of bruises, but often she was unsure how she'd acquired them.

To her the trance was entirely otherworldly. She said it began "as if something were passing over my brain making it numb; a sensation similar to that I experienced when I was etherized, only the unpleasant odor of ether is absent." She said the room began to chill and the people in it to shrink. Light faded until the room was black. When she woke, her hands and arms tingled. She could see the occupants of the room again, but only at a distance. She often emerged exclaiming, "Oh, how black you are!"

Her trance personality, the impossible Phinuit, was, Hodgson judged, obviously anything but what "he" claimed. Hodgson had never found evidence that the French doctor existed in reality. After Phinuit claimed to have been born in Marseilles, Hodgson had persuaded Charles Richet to go through birth records in the port city. Richet found nothing. Hodgson—along with Richet and James—declared the so-called doctor a secondary personality of Mrs. Piper.

The consensus was that Phinuit was a coping device, a subconscious way for the medium to protect herself against whatever mental battering took place in her trance state. As a fictional character and a mental buffer, the doctor made a kind of strange sense. As a spirit, he made no sense at all. But then, neither Hodgson nor James was convinced that Mrs. Piper did commune with spirits. They had yet to find what would qualify as undisputed proof of that communication.

Out of her trances came extraordinary personal insight. But it was muddled, tangled with vaguely Christian notions of life after death, ambiguous messages of cheery goodwill, and rather pointless conversation. The theory "of spirit control, is hard to reconcile with the extreme triviality of most of the communications," James complained. "What real spirit, at last able to revisit his wife on this earth, but would find something better to say than that she had changed the place of his photograph?"

For the real believers, the positive note came in the report's rather cryptic close, promising that other results would be forthcoming. "Mrs. Piper has given some sittings very recently which materially strengthen the evidence for existence of some faculty that goes beyond thought-transference from the sitters, and which certainly [on its face] appears to render some form of the 'spiritistic' hypothesis more probable."

Hodgson, it seemed, had chosen to end his report on a cliffhanger.

"BETWEEN THE DEATHS and apparitions of the dying person and the living a connexion exists which is not due to chance alone. This we hold as a proved fact. The discussion of its full implications cannot be attempted in this paper; nor, perhaps, exhausted in this age."

Writing late and early, on gorgeous June mornings and damp July afternoons, Nora had finished the report of the Census of Hallucinations. Sidgwick planned to give a preliminary presentation at the International Congress of Experimental Psychology in August 1892.

Six countries had participated—England, France, Germany, Russia, Brazil, and the United States—and each group's findings had confirmed the others' work. The British census was the largest—17,000 surveyed—and the American study, coordinated by James and Hodgson, was second, with

7,123. All concluded that death-day apparitions occurred in startling numbers. The American survey found that these "ghosts" occurred at 487 times
the rate predicted by chance. The British calculation was 442.6 times chance.

Both groups used a statistical formula worked out by Nora. Employing
the British figures from the Registrar-General, she calculated that the
chances of any one person dying on a given day were 1 in 19,000. The possibility of a given single event, such as a recognizable "hallucination" of a
certain person, occurring on that same day was also one in 19,000. So for
every 19,000 deaths, there should be only one such occurrence.

Out of their 17,000 respondents, 2,272 had claimed to have seen such
an apparition near the time of death. The British SPR had winnowed this
down to 1,300 by removing all reports in which dreams or delirium were
noted. They then winnowed further to a narrow window of time. Only
eighty of these sightings had occurred within twelve hours of death. They
winnowed again, throwing out all cases in which there was chance of prior
knowledge of when the death was expected, such as that of an elderly or
ailing relative. They then removed from their list any ghost story that relied
on only one person's claim.

Thirty-two cases were left, and by Nora's calculation, that was a big
number. At a rate of 1 out of 19,000, they should have seen .0723 instances
among the 17,000 surveyed. Instead, they had 32 cases with solid evidence
behind them, which was 442.6 times the chance rate of .0723. Hodgson
had used the same statistical process to arrive at the American numbers.

Despite that consistency, James worried that the numbers remained
unconvincing. He felt dissatisfied with the U.S. census. It lacked the
numerical weight of the British one; as he wrote to Nora, "our census has
been a terribly slouchy piece of work and comparing it with yours makes
me blush throughout." He blamed himself; he'd paid scant attention to the
weight of the stories until late in the research. Only while reading and veri-
fying the reports had he discovered—as Edmund Gurney had done during
the work on *Phantasms*—the compelling pattern they contained. By that
time, he confessed to Nora, he had let the correspondence fall in arrears
and run out of time to corroborate many of the cases.

James wished they'd been able to round up the 50,000 recommended
by Gurney, to overrun their critics with the power of their numbers. But

maybe such a coup wasn't vitally important to the cause. Maybe a single census, no matter how substantial or indisputable, would never win credibility for the SPR. Perhaps it was only the slow and gradual process of survey upon survey, census upon census, which could win over the doubters.

"I never believed and do not now believe that these figures will ever conquer disbelief," James wrote in a letter to Henry Sidgwick that accompanied the American report. "They are only useful to rebut the dogmatic assurance of the scientists that the death warnings are chance coincidences." The psychical researchers would conquer disbelief when they could also reveal what those death warnings, those unexpected apparitions, really meant in terms of life and death and Huxley's unknowable God.

8

THE INVENTION
OF ECTOPLASM

THE WINTER OF 1892 howled across the Atlantic coast of North America like an ill-tempered spirit, spitting snow across the landscape. In hard-hit New York, where a seemingly perpetual crystalline haze veiled the air, horses struggled along Fifth Avenue, heads down against the wind, wading through a treacherous mire of slush over ice.

George Pellew, a thirty-two-year-old philosophy student and writer, was among that season's many victims. He was riding along an icy path in Central Park one bleak February day when his horse lost its footing. Pellew died in the resulting tumbling fall.

Dick Hodgson came down from Boston for the funeral, mourning another friend gone too young. The Australian had been on a lecture visit to New York when he'd met Pellew, an outspoken skeptic on the subject of psychical research. Hodgson enjoyed a good argument, and they'd struck up a friendship, as much because of Pellew's prickly stance as in spite of it. On his subsequent visits to the city, the pair would meet for beer and talk, occupying a tavern table for hours while they debated immortality and the odds of life after death.

The prospect of floating around after death as some ill-defined energy field or specter seemed to Pellew an unlikely idea, even a ludicrous one. Hodgson had agreed, to a point. He was willing to concede that spirit life was improbable, yes, but not impossible.

A few months before his death, Pellew had made a half-joking promise. If Hodgson was right, Pellew was willing to prove it. If he died first, he would return and "make things lively." He would make himself so obvious, Pellew threatened cheerfully, that his friends wouldn't be able to deny him.

Hodgson had laughed.

THE BITTER FEBRUARY gave way to a bitter March. Then, five weeks after the fatal accident in Manhattan, a new voice interrupted one of Mrs. Piper's trances. The personality identified itself as George Pellew. Soon, and persistently, this new presence would alter the very nature of a Piper sitting.

Although G. P.—as Hodgson came to call the personality—manifested himself at first as a voice, he preferred to communicate through automatic writing. Early on came a few bizarre and hectic trances in which Phinuit answered one question verbally, while the medium's right hand wrote G.P.'s answer to another on a paper tablet. Gradually, though, the familiar Phinuit began to fall silent. When Hodgson asked a question, the medium's only response would be the scratchy sound of her pencil (pens were never used because of the need to continually dip them in ink) moving across a page.

It was G.P.'s arrival that provided the concluding note of optimism to Hodgson's report on Mrs. Piper. Hodgson wasn't convinced that this new personality was a spirit. Perhaps it was no more than another peculiarity of Leonora Piper's subconscious. But unlike the dubious Dr. Phinuit, G.P. claimed to be someone Hodgson knew personally. That fact offered a realm of possible tests to determine who or what—if anyone or anything—was communicating through the baffling medium.

Hodgson began by making a list of old friends and family members of the dead writer. He would invite them, as many as would agree, to come anonymously and check their knowledge against that of the trance personality. Maybe they would confirm that this new spirit guide really was George Pellew. Maybe they would not. As always, the investigative strategy was as interesting to Hodgson as the possible results.

His investigation of the so-called ghost of George Pellew was based upon a simple idea, with a twist. He would bring more than a hundred visitors, eventually, to sit with Mrs. Piper. Some would be friends of the dead man; some would be strangers to him. But she would be given no relationship clues. No participants would be allowed to tell their names or whether they had any connection to G.P. They would be allowed to improvise personal tests, but they would not be allowed to give any explanation for them.

One visitor brought a photograph of a building.

"Do you recognize this?"

"Yes, it is your summer house."

Which it was.

Another woman placed a book on the medium's head.

"Do you recognize this?" she said to G.P.

"My French lyrics," he answered.

That was right too.

Another visitor, a man, simply asked, "Tell me something, in our past, that you and I alone know."

As he spoke, Mrs. Piper sat slumped forward into a pile of pillows on the table, her left hand dangling limply over the edge, her right hand coiled loosely around a pencil. Next to her right side, a pad of paper sat on the table. Suddenly, her fingers tightened and she began to write, wildly, filling pages, ripping them off, thrusting them away from her.

Hodgson moved to the other side of the room. The man began flipping through the pages. He paled and folded the papers. They were too private to read aloud, he told Hodgson.

But he was "perfectly satisfied, perfectly."

"I COULD NOT distinguish anything at first," G.P. told a friend during one of the sittings. "Darkest hours just before dawn, you know that, Jim. I was puzzled, confused."

"Weren't you surprised to find yourself still living?" his friend asked in return.

"Perfectly so. It was beyond my reasoning powers. Now it is as clear to me as daylight."

.   .   .

IT WAS IN the summer of 1893, while still traveling abroad, that William James received an unexpected letter from a colleague at Harvard, a researcher who'd decided to sneak a visit to Leonora Piper, Boston's most famous medium. The professor had contacted Hodgson using a fake name. Even after the sitting, he'd not offered his real one. Mentally, he'd been snickering as the medium slumped into her trance, as her hand began to write.

"I asked her barely a question, but she ran on for three-quarters of an hour, telling me names, places, events, in the most startling manner." Someday, he promised he would tell James what she had revealed; for now, he'd just say it was information not meant to be shared.

Still, there were a few interesting details that he wanted to pass along. Once again, Mrs. Piper had revealed her peculiar psychometric gift, as if she could read a story from a material object. It made no physical sense, but there it was:

The professor had brought a single circle of gold, one that once belonged to his dead mother. The ring had been one of two, a set that he and his mother had exchanged one Christmas.

Each ring had been engraved with the first word of the recipient's favorite proverb. Long ago, he'd lost the one she'd given him. But the previous year, when his mother died, the ring he'd given to her had been returned to him.

The professor was holding that ring in his hand during the sitting, hiding the word as he inquired, "What was written in Mamma's ring?"

"I had hardly got the words from my mouth till she slapped down the word on *the other ring*—the one Mamma had given me, and which had been lost years ago.

"As the word was a peculiar one, doubtfully ever written in any ring before, and as she wrote it in such a flash—it was surely curious."

As an educated man, a scientist, no believer in the silly afterlife ideas of the spiritualists, the professor would admit only to being curious, as he explained carefully to James.

. . .

FRED MYERS AND OLIVER LODGE were coming to America. They were to present the latest SPR research at the Congress of Psychical Research. Both were looking forward to visiting Chicago in August 1893, especially because the city was hosting the newly opened World's Columbian Exposition. Myers's only disappointment was that, despite his pleas, James refused to cut short his sabbatical to join them, even with the added lure of the new World's Fair.

"Your letter rec'd, bristling as usual with 'points' and applications," James wrote back to Myers, from Italy where the James family had happily settled for the summer. He was unmoved by Myers's persuasions. He had no new evidence to present. Further, as he warned Myers, in the case of their controversial line of work, James worried about making too vigorous a push too soon. "What we want is facts, not popular papers, it seems to me, and until the facts thicken, papers may do more harm than good."

As a compromise, though, James had asked Hodgson to give a brief report on his Piper results at the convention. There were so few good mediums—at least, he and his fellow investigators tended to eliminate most as fraudulent—that Mrs. Piper now stood out like the only flower left in a denuded garden. That was one of the issues Hodgson wanted to raise—the need to find mediums before they sold out to the demands of the profession.

As it turned out, Myers and Lodge were so fascinated by the World's Columbian Exposition that Hodgson—after making his speech—found himself dragged across the white-marbled landscape of the fair, from the huge, glass-walled Fisheries Building to the tiny Chinese pavilion, with its heavy furniture carved with the shapes of lazy, coiling dragons, to the glittering technology exhibits. Hodgson felt that he'd seen "nearly all of the World's Fair that I care to," as he wrote to Mrs. Piper.

He missed her, he said. He also missed being in the thick of work. He'd been hearing from his associates that the G.P. sittings were still going astonishingly well. They were much more interesting to him than the marvels of the exposition. He couldn't wait to get back.

. . .

THE JAMES FAMILY returned to Boston in early September 1893, newly invigorated by their lingering summer in Italy. The very air of Cambridge, the brick and stone of Harvard, seemed "surcharged with vitality," James declared enthusiastically.

It didn't take long for that sense of euphoria to wash away. Despite the sizzling pace of technological advances—kinetoscope movies were playing in New York, Henry Ford had just built his first car—America was stumbling through an economic recession. Five hundred banks and thousands of businesses had failed over the summer. James was forced to sell stock to meet living expenses. He and Alice talked of selling their big house on Irving Street in Cambridge.

In Italy, he'd started dreaming of early retirement. He could feel that fantasy evaporate along with his good mood. His profession of psychology seemed "paltry and insignificant," his colleagues mired in laundry lists of laboratory details. And psychical research seemed too dependent on a single, if phenomenally baffling individual, Mrs. Piper.

James worried that no other medium offered comparable results. Even if there were another Mrs. Piper somewhere, the psychical research group had not found her, or him. They'd hardly looked. The very few dedicated investigators were preoccupied with other, more substantially rewarding work. He worried about Mrs. Piper, too. The SPR had agreed to pay her $200 a year to guarantee that she accept no fee from any other source. She was earning her money, James acknowledged, but the very act of taking payment sullied her in his estimation. Her character seemed weaker to him, he told Hodgson.

The general character of academic scientists displeased him as well. He'd invited eight colleagues from Harvard to observe Mrs. Piper, telling them of recent extremely positive developments. Five refused, one informing James that even if something happened, he wouldn't believe it. "So runs the world away!" James wrote. "I should not indulge in the personality and triviality of such anecdotes were it not that they paint the temper of our time."

The return to the United States had soured James's outlook—not just toward his American colleagues and peers but toward the rest of the world as well. Sometimes he couldn't help but wish that all of them—the psychi-cal researchers and their opponents, the prejudicial naysayers—would just go away and leave him alone.

He was still in that funk a month later when a letter arrived from the irrepressible Fred Myers, inviting William James to become president of the British Society of Psychical Research. Invitation was hardly the word. The letter was more in the nature of a summons. "The *first* reason is, of course, your position as a psychologist," Myers explained. All of the SPR's most interesting scientific ideas—telepathy, subconscious communi-cation, the mental state of mediums—were rooted in the new science of psychology.

But his being an American was good, too. Myers felt that James's nationality would emphasize the society's trans-Atlantic inclusiveness. The Sidgwicks agreed, he said, and would even waive the requirement that James attend the London meetings; he would simply appoint someone to read any musings from the president. "We trust that you will grant our prayer. We cannot see that it will hurt you; and we see very clearly that it will *help* us."

The timing couldn't have been worse for James. Didn't he have enough to trouble him, what with accumulating bills and other work, and col-leagues who continued to annoy him? He fired off a hasty refusal, blaming his usual assortment of physical ailments.

Such excuses carried no weight with Myers. He replied that he was always sorry when friends were not in top form, but it seemed to him that James's ill health was never perceptible to anyone but William James. Cer-tainly, he, Fred Myers, saw no evidence of frailty. What he saw instead was a rare man, full of wisdom and delight, the perfect SPR president, held back by only one character flaw.

That would be a tendency to give up too quickly. "It seems to me you lack one touch more of *doggedness* which would render you of even more helpfulness in the world than you are."

If his friend were really ill, Myers conceded, he would agree to let him defer his acceptance. That would be a shame, he added, since Myers thought the job could actually improve James's health and outlook: "Mrs. Piper is all right—and the universe is all right—and people will soon pay more money to the SPR—and an eternity of happiness and glory awaits you."

On December 17, 1893, a two-word telegram whistled its way across the Atlantic. Myers read it with gratification, but not surprise. It said only: "James accepts."

IT HAD TAKEN cajoling, pleading, and threatening—even for the Master of Turin—but Cesare Lombroso had launched not just one, but a series of scientific investigations of Eusapia Palladino.

The first was something of a shambles. Led by Lombroso, the team also included a Russian psychologist, two Italian physicists, and the French physiologist Charles Richet. The sittings took place in the private home of a helpful Milan resident, who had agreed to have his house searched in advance and his parlor, to be used for the sittings, locked and sealed after each test.

But as Richet would tell his SPR colleagues, Lombroso was now so enthralled, the other scientists so unnerved by confronting a medium, and Eusapia so prone to scream like a fishwife when she didn't get her way that they lost control of the experiments almost immediately. The scientists wanted full light. She insisted on a dim red light in a darkened room, claiming that bright illumination would put off the spirits. They gave in. The researchers asked her to stand, instead of sitting at the tables she planned to levitate, so that her feet were not concealed. She refused, declaring that her legs and knees trembled so violently during levitations that she could not possibly stay upright. They gave in again. And trying to control her hands and feet was like wrestling with a freshly caught squid. She was never still, Richet complained, always twitching her fingers away, wriggling her toes.

Most of the time, it was impossible to be sure that she wasn't sneaking a hand away to produce a phantom touch, or nudging furniture with her feet or knees. Most of the time, Richet knew he was observing some rather obvious cheating. But every once in a while, the whole feel of the sittings changed: the sneaky medium disappeared, and a pale, still woman replaced her; the curtains began to shiver, as Lombroso had reported earlier, billowing in that nonexistent breeze. Hurrying to open the draperies, Richet would have sworn that he felt the touch of cold hands, although that could have been his nerves. No one was there, no wire, no body, no anything except empty air between the curtain and the window.

He could explain away the common cheat. It was the other, more elusive

Eusapia who bothered him, the one who sat pinned to her chair while the cold fingers of the supernatural seemed to crawl into the room. As James had complained, there were few real mediums available to psychical researchers. It occurred to Richet that, with patience, Eusapia might offer a chance to study the difference between what was real—and what was contrived.

He tested her again, without the Italian and Russian scientists, who, he thought, had compromised the earlier observations. Those experiments yielded the same frustrating mix of deliberate fraud and inexplicable event.

In one sitting, at the Psychological Institute in Paris, he'd brought in several witnesses, including the formidable physicist Marie Curie, who Richet hoped could tell the other observers if there was any sign of unusual energy in the room. He and Mme Curie sat on either side of the medium, each gripping one of Eusapia's hands. "We saw the curtain swell out as if pushed by some large object," he noted. Richet reached up and grabbed the bump behind the fabric. It felt like a hand, but one with sausagelike fingers, much bigger than Eusapia's "little hand," and with nothing beyond the wrist itself. He glanced back to make sure the medium's hands were still secured. Mme Curie assured him that she'd kept an unbreakable clasp on Eusapia's fingers.

Richet tried another experiment, laying pieces of smoked paper on a table some distance away from the medium. Pale hands appeared and pressed against the paper. When he picked up the paper, the dark film of smoke had worn off in places, as if a finger had been rubbing at it. Eusapia's hands remained clean, untouched by smoky residue. Those creeping hands, what to do with them? How to define them?

Out of his growing frustration, Richet invented a new word for the phenomena—*ectoplasm,* cobbled together from the Greek *ecto,* "exterior," and *plasm,* "substance." "C'est absolument absurde, mais c'est vrai!" Richet exclaimed, deciding, like Lombroso before him, that he required reinforcements.

RICHET'S FAMILY owned a tiny island in the Mediterranean, just off the French Riviera, tucked amid three famously beautiful islands called the Isles of Gold. Ile Roubaud was a rocky scrap of land, scrubbed by light and polished by water. Only three buildings occupied the island. The Richet family summer cottage, though not large, bristled with towers, turrets, verandas, and porches. Nearby stood a lighthouse and a simple cottage for the lighthouse keeper.

The simplest route there was to sail through the French government's salt lagoons, where the dried layers of sea salt were prepared for sale. The salt ponds gave the voyage to Ile Roubaud a slightly unreal feeling, a passage through a landscape of almost blinding white where the sun dazzled on the crystal layers rimming the water.

Beyond, across a blue sparkle of sea, lay the small island, encircled by pinwheeling seabirds. Richet thought it the perfect place to run tests on a troublesome medium. He would search her, isolate her on the island, and then he would see what happened to her so-called powers in the luminous light of Ile Roubaud.

In the summer of 1894, Richet invited a small party to join him for the grand experiment. His guests—and witnesses and collaborators—were Fred Myers, Oliver Lodge, the Polish psychologist Julien Ochorowicz, and Richet's personal secretary, who would be there to take notes.

Lodge remembered the journey to Richet's island as something of a comedy of errors. Traveling by rail to the Mediterranean coast, he and Myers disembarked at one station to have a quick drink. The train left without them. They caught up with it in Avignon. But Myers wouldn't leave that ancient city without touring the Palace of the Popes and the famous crumbling bridges over the Rhône. Finally they reached the coastal spa city of Hyeres. With some difficulty, the pair hired a boat and made their way past the white glimmer of the salt lagoons, guarded by soldiers installed to protect the government's lucrative monopoly on salt sales, apparently bent on preventing the theft of a single white crystal.

"The salt monopoly has curious results," Lodge noted in his diary. "It appeared that the peasantry were forbidden to take a bucket of water out of the sea." The French soldiers returned the stares of the voyaging Englishmen; the air was filled with the dry creak of seagull voices overhead. Lodge feared that this was going to be a very odd visit.

In that, he would be proved absolutely correct.

THE INVESTIGATORS wedged themselves like tinned sardines into the family cottage. Lodge shared a bedroom with Richet. Myers occupied a child's bedroom and slept folded into a very small bed, which he claimed to share with a family of flies. Ochorowicz bunked on a balcony. Lodge traded places with Ochorowicz for a while, but "found the only other occupant was a mosquito, who woke me up punctually at five every morning." Eusapia had a room to herself, a respectable distance from the others. Richet's secretary commuted in from Hyeres for the tests.

The summer heat parboiled the island. The air steamed around them. Most afternoons, Richet sought escape in his small boat, trolling for fish and the elusive breeze dancing over the cooler waters. Lodge and Myers shed their usual sober dress and spent the days in their cotton pajamas, clambering over the rocks and occasionally dousing themselves in the sea.

There was nothing else to do, really. The sittings were held in the evenings, and the researchers had been instructed not to socialize beforehand with their captive medium. The group did gather for dinner, cooked and served by the lighthouse keeper's wife. These meals tended to lapse into cacophony. The investigators conversed in French. Eusapia spoke Italian, in the Neapolitan dialect, loudly. She liked to shout down her companions, demanding that they listen to her life stories, over and over.

She particularly liked to recall the dramas of her life, and she not only verbally re-created the moments, she acted them out in style. When she told of the brigands of her childhood—thieves who had reportedly killed her parents—she did it by leaping onto the table and, in Lodge's fascinated words, "waving kitchen knives about like sabers."

It was worse if the conversation turned to her occult abilities, especially if any shard of doubt appeared. She routinely worked herself into a screaming fit—"a Neapolitan rage," as Lodge described it—when the subject of trickery arose. The problem was that it always arose, because she always cheated when she could. Ochorowicz called this "reflex fraud." He suspected that it was a game for her, almost a flirtation; she liked to see what she could get away with. Only after Eusapia tested the limits would she settle into the business of being a medium.

Ochorowicz was a big, fair-haired man with a deceptively relaxed demeanor. He was suspicious enough to search Eusapia, her luggage, and her room regularly and without warning. She put up with it, but she resented it. Did these scientists think the spirits complied, the power appeared, every time? She liked to get the job done. If she couldn't call upon super-natural powers, she would use her own.

As she told Ochorowicz, it was up to the investigators to discern which occasions were which.

Eusapia and her inquisitors gathered around a lamplit table in the early evenings. Each night, one of them—usually Richet's secretary—took notes. The others took turns holding the medium's hands and feet and even her head. Sometimes they used a traplike device designed by Ochorowicz, which caged her feet and caused a bell to ring if she moved them.

Lodge and Myers found the medium just as puzzling as Richet had warned them she would be. A music box sitting on the table began to play and then rose to press against Myers's chest. When it dropped to the floor, Myers stumbled forward. Something was pushing him from behind, he called out; would they please go look at it? His colleagues could hear a slap-ping sound against Myers's back, but there was nothing there. A white pro-tuberance suddenly extended from the medium, stretching in the dim light until it prodded Myers in the chest. He flinched back; it was fingerless, but it felt, he said, like a hand grasping his ribs.

There were nights when a brass key sailed off the library table to fit itself into a door lock. Other times, a strange yellow and blue glimmer of light winked on, off, on again in the empty air. And there was that strange wind, rising out of a vacant corner, stirring the edges of the room.

"There is *no doubt* to this business," Myers wrote to James, "& we are plunged into the grossest superstition."

MYERS AND LODGE left so convinced of Eusapia's legitimacy that they determined to write up a report on her and submit it to a science journal. Their first choice was *Science* magazine. Dismayed, James did his best to discourage that plan, reminding them of the fate that William Crookes had suffered at the hands of female mediums. The Sidgwicks were equally unenthusiastic; they'd also sat with Eusapia and found her puzzling but not persuasive.

The Palladino situation was a "crisis," Sidgwick wrote to a friend. The SPR had worked hard for "a reputation for *comparative* sanity and intelli-gence by detecting and exposing the frauds of mediums." Sidgwick hated to see that credibility squandered on a medium whose phenomena—levitations and ghost hands—were the mainstay of fraudulent mediums, many of whom had been exposed by his organization. He worried equally about losing the appearance of objectivity: "It will be rather a sharp turn in our public career if our most representative men come forward as believers."

Consider the reputation of Cesare Lombroso. Still struggling to explain his encounters with Eusapia, Lombroso had now published a theory that the medium could access an unknown psychic force, capable of reaching into the "ether" for its power. In their search for scientific acceptance, the members of the SPR definitely did not want to find themselves associated with ether theories of psychical powers.

The most misbegotten spiritualists, Helena Blavatsky among them, had embraced this idea of the "ether." The ether, being invisible, was difficult to describe, but it was postulated as a kind of cosmic cream that oozed through space, pervaded all matter, and acted as a kind of spiritual filler between this world and others. Blavatsky claimed that the breath of early gods had formed it, and that only the most advanced mystics (including herself) could use it to acquire unearthly, etheric powers.

The spiritual version rested on an idea from the science of physics, dating back to Isaac Newton, that an unknown material might fill unoccupied space and serve as a conducting medium for light and heat. Oliver Lodge was among those who'd looked for evidence of such mysterious matter, and he'd also helped discredit the idea of Madame Blavatsky's kind of spiritual ether. The previous year a British shipbuilder and spiritualist had provided money for Lodge to design a machine specifically to test for the psychic ether.

Lodge's machine used a pair of powerful dynamos to spin metal disks into a blur, whirling them 4,000 revolutions per minute, producing a sizzling electrical charge. Its purpose was to test a favorite theory of the spirit believers, that the ether had a natural affinity for charged atoms, that its ability to carry electrical energy might be thus responsible for psychic powers. The machine contained instruments to measure any "etheric" effect on the sparking, electrically charged disks.

The instruments registered absolutely nothing.

Resurrecting the ether was likely to make them all look like fools, James wrote to Myers; already it made Lombroso appear as "the greatest donkey of the age." Prematurely advocating for Eusapia Palladino, without thoroughly eliminating all the suspect parts of her séances, would be a mistake. The SPR didn't really understand her—any more than Lombroso did—and until then, James urged self-restraint. "You ask what I think about popular publication? I must confess myself extremely averse," he wrote to Lodge. "The more startling the secrets we have to disclose, the more, in my opinion, should we calmly pursue the tenor of our ways and publish proceedings at their due date. The stuff will keep and the bigger the bomb to be exploded at once in the proceedings, the greater the shock."

IN THE WARM AUTUMN of 1894, Dick Hodgson was running like a dynamo himself, hunting ghost stories, reading journals, writing letters, speaking, holding meetings, and writing up the G.P. sittings for his next Piper report.

He thrived on the work, along with what recreation he could sneak into his days: handball and pool at the Union Boat Club, drinking beer with friends, hurrying through his favorite bachelor meals of eggs, bread, and tea. His letters to his friend Jimmy Hackett rang with optimism; psychical research looked more promising than it had ever seemed before.

The news that Myers and Lodge wanted to gamble the reputation of the whole enterprise on a rather shady medium ruined that sense of well-being. Hodgson understood that everyone wanted a base of psychics that extended beyond Leonora Piper. But they were fooling themselves if they thought that this woman would ever be credible.

He immediately cabled to Myers, imploring him to back away from Eusapia Palladino. Myers was beginning to feel beleaguered. His reply had a snap to it. Hodgson hadn't been there, Myers wrote back, so the ASPR secretary couldn't properly evaluate what had happened.

Now a little angry, Hodgson stayed up late for several nights running to write a crisp—and undiplomatic—analysis of how the Ile Roubaud group had been deceived. They might think they could see through Eusapia's little cheats to a bigger truth. But in fact, Hodgson concluded, that was upper-class arrogance. She might not be Cambridge educated, but she was far more cunning than they.

He mailed his report to Nora Sidgwick, along with another tirade in the form of a letter. He dismissed Lodge as a poor investigator, despite training in physics, and Myers as easily gulled once a medium had gained his sympathies. "This is part and parcel of his big, poetic, divine genuine soul, & he can't help it!"

Nora was then editor of the *Journal of the Society for Psychical Research.* As ever, she was unmoved by emotional appeals. But cold intelligence called to her like a siren song, and Hodgson's dissection of the Ile Roubaud experiments was both clinical and smart. She published his analysis, unabridged, in the April 1895 edition. She expected to infuriate every person who had been on that island, and frankly, she didn't care.

Hodgson was ruthlessly to the point: Eusapia behaved like a fraud because she was one. By constantly twitching her hands away and moving them, she could lead two sitters to grab the same hand, freeing her other one. Or she might even trick them into holding each other's hands. In the dim room they'd allowed her, the investigators couldn't see what they were holding anyway. Their own less than meticulous records, Hodgson added, contained no proof that they'd effectively prevented her from using such methods. Thus, all their mysterious little stories were meaningless.

Hodgson knew all about medium tricks, and he was happy to share his knowledge in detail: once Eusapia gained even a little mobility, she could levitate tables with secretly attached hooks, move objects, and prod researchers with a collapsible steel rod attached to her knee, all the standard devices of the trade. Those pale hands and projections were probably molded paraffin. She could have rigged the room when they weren't looking, stringing it with fine twine that would not be visible in the dim light.

And the "inexplicable" wind? Another well-known trick. An inflatable bladder, a balloonlike device that puffed air when compressed. All she had to do was sit on it. As for ectoplasm or ether, please. The very concepts made him want to laugh. Hodgson only wished that his colleagues would start behaving like adult investigators.

As Nora had foreseen, indignation ran at high tide among the Ile Roubaud investigators. One after another they hurried forward to demolish Hodgson's case against them.

Richet pointed out that they *had* searched Palladino, caged her feet, and conducted experiments in light as well as dark. They weren't so dumb or so blind that they wouldn't have noticed a rod stretched across the room, Richet said; they probably would have fallen over it when they hurried to investigate a moving object.

Not only that, Myers wrote, Eusapia didn't wiggle away quite so often as Hodgson implied. Many times they'd had her hands securely locked in theirs. And did Hodgson really think that Myers wouldn't notice a hand switch, wouldn't realize he was suddenly holding Lodge's "massive, steady, round-nailed hand" instead of Eusapia's "small, perspiring, quivering, sharp-nailed hand"?

Ochorowicz wrote that Hodgson, not a trained scientist, had missed the point. The most interesting possibility offered by physical mediums such as Eusapia was the rare, occasional hint of a different kind of power, that of telekinesis. Like *telepathy,* this was a new word, patched together from the Greek words for "far" and "movement."

After all, Ochorowicz continued, if one medium could extract information from an object—as even Hodgson acknowledged Mrs. Piper seemed able to do—then perhaps another could exert energy on objects, not pulling facts from them but buffeting them with energy, making them move in response.

He cited a particular incident to illustrate that possibility: During one of the sittings, they had wedged Eusapia between Myers and Lodge, shoulders against shoulders, each man gripping her hands. Ochorowicz had then placed himself underneath the table to clasp her feet. A large table, some four feet away, had lifted into the air and turned itself upside down. They'd weighed it afterward. It was a hefty forty-eight pounds, suggesting that a tug on fine threads, had there been any such devices attached, would have accomplished nothing except to snap them.

It was possible, he continued, that Eusapia didn't know her own gift. Responding to the spiritualist claims of the time, influenced by the very people who studied her, she liked to credit spirit power for her phenomena. But it might be nothing of the kind. It might be a power or energy form that she herself did not understand.

Certainly, he and his fellow investigators didn't understand it. And from the vantage point of his Boston office, Ochorowicz wrote, Hodgson couldn't possibly understand it either.

Faced with such divisiveness in the ranks, bombarded by angry and opposite opinions, Henry Sidgwick made a judicial decision. The investigators would simply start over.

This time, they would test Eusapia on their terrain. The British SPR decided to bring the Italian medium to England, as they had with Mrs. Piper. Further, they would train for her visit. They began practice sittings at the Myers's house in Cambridge, trying different ways of holding hands and feet. Myers was particularly impressed—and secretly amused—by Henry Sidgwick's newfound skill for dropping to the floor, his white beard trailing over the carpet, while he anchored Nora's feet in place.

Hodgson wasn't impressed all. He fired off an immediate demand that they bring him to England as well. He didn't mean to be rude, but he was sure that without him they would get it wrong again.

EUSAPIA HATED CAMBRIDGE.

Everything was cold—the climate, even in this so-called summer of 1895, the oh-so-polite conversation, and the self-contained British personalities. She was a warm-blooded woman, hot in nature. In middle age, she was discovering that the occult could really steam her up.

She tended to wake from trances hot, sweaty, and, well, aroused. Several times, she'd tried climbing into the laps of the male sitters at the table. In England, the men had a distressing tendency to stand up in response, rather than take advantage of the opportunity.

They wanted her to be comfortable so that she would be in a receptive state of mind for the séances. Evie Myers took Eusapia shopping, allowed her to cook Italian meals in their kitchen, listened smiling to all the medium's chatter, although Evie herself spoke only a few words of Italian and had no idea what Eusapia was talking about. Evie also photographed the stubby Eusapia, who demanded to be draped in Sidgwick's austere academic cap and gown.

"Sidgwick has to *flirt* with her," Myers wrote to James, but he begged James to keep that part a secret. "This is not for Philistine ears." Nora, serene as ever, came by regularly to translate for Evie and to write letters home in Italian for Eusapia, who had never learned to write.

The Myerses' eight-year-old son, Leo, was recruited to play games of croquet with the medium. She enjoyed it, standing on the smooth green lawn, catching pale rays of northern sunlight, slamming around the bright-colored balls. But Leo complained to his parents that she cheated every time.

Nevertheless, Eusapia fell into an ill-tempered sulk, which carried over into the sittings. Exhibiting an indifference toward the whole enterprise, she refused to be tied in place, sometimes wouldn't allow her feet to be held, yanked her hands away from confining grips. Little happened. Once or twice a table tipped. A few trinkets skittered across a mantel.

The most interesting result occurred during a visit from Lord Rayleigh, who had brought with him a friend and fellow physicist, J. J. Thomson. The tall thin Thomson and shorter, stockier Lord Rayleigh made a remarkably good pair of observers. Both men would win Nobel Prizes in physics within the following decade, Rayleigh for his work in atmospheric chemistry and Thomson for his elucidation of atomic structure.

At this moment in Cambridge, though, they were sitting in the Myerses' library, watching Eusapia with ironic detachment. As they sat, suddenly the curtains billowed out before a closed window. Thomson went over to measure; the fabric had blown out two and one-half feet, by his calculation. The medium sat some feet away, eyes shut, a faint frown on her face.

Rayleigh walked over and put his hands against the curtains. They pushed back against him. He put his hand between Eusapia's back and the curtain, felt along the floor between, and found no device, nothing to connect the cloth to the medium. He and Thomson weren't willing to call it supernatural. They would commit to calling it odd—but as Rayleigh admitted, "odd" didn't do it justice.

HODGSON ARRIVED AT Myers's house with all the appearance of a born mark. Not a trace of the brusque and suspicious investigator was in evidence. Suddenly, he was a little clumsy, a little dumb, and uncharacteristically gullible.

Eyes wide, he shambled into the séances. He sat next to Eusapia, holding her hand, but not too tightly. He allowed himself to be distracted, jumping at shadows, watching others in the room.

It was open invitation to cheat—and she took it.

Within a single séance, he'd seen her wiggle a hand away, spread the other hand so that two hand-holders were each gripping an edge of that single hand, manipulate objects with her freed fingers. With her feet barely held, she used them as well, kicking the table, moving chairs. When her hands were securely held, Hodgson watched her drop her head onto a sitter's shoulder, listened to the man's startled exclamation that a hand had touched him in the dark.

As Hodgson had said before, and now said again, she was an obvious cheat. She was so easy to catch that she wasn't worth any more of his time. He was going back to Boston, and he was done with the British SPR's efforts to make something out of nothing. He didn't want to hear anything more about bulging curtains, inexplicable winds, eerie white hands—or Eusapia Palladino.

As the year 1895 moved forward, civilized behavior—as James noted ruefully—seemed to be rapidly going in the opposition direction.

Italy was fighting in Abyssinia; the Chinese and Japanese were battling over the island of Formosa (later called Taiwan); in Cuba, citizens had risen in an attempt to shake off Spanish rule. The United States was quarreling with Britain over colonies in South America. The question of the precise border between Venezuela and British Guiana had become so heated that President Grover Cleveland threatened war against the United Kingdom.

"Well, our countries will soon be soaked in each other's blood," wrote James to Myers. "You will be disemboweling me, and Hodgson cleaving Lodge's skull." Joking aside, James loathed Cleveland, whom he considered a posturing hothead, a leader too impatient with diplomacy, too eager to spill someone else's blood. "All true patriots here have had a hell of a time," James complained, feeling that support of Cleveland's warmongering didn't really speak well of his countrymen either. "It has been a most instructive thing for a dispassionate student of history to see how near the surface in all of us the fighting instinct lies."

The joke about Hodgson braining Lodge was also, unfortunately, rather too close to reality. The Cambridge sittings had fractured the usual sense of unity among psychical researchers, diluting their comfort in being a small band of Davids working to overcome an army of scientific Goliaths.

Henry Sidgwick had taken Hodgson's position and refused to publish the few positive observations from the Eusapia sittings. With Nora's support, Sidgwick declared that the woman was an obvious fake, and he was weary of giving her free publicity. "It has not been the practice of the SPR to direct attention to the performances of any so-called 'medium' who has been proved guilty of systematic fraud," he wrote. "In accordance, therefore, with our established custom, I propose to ignore her performances for the future, as I ignore those of other persons engaged in the same mischievous trade."

Myers and Lodge remained angry with Hodgson, who they thought had destroyed any chance of decent experimental work. They hadn't in the least appreciated his gullible act. Hodgson had deliberately baited a trap, knowing that Eusapia always cheated if given the opportunity, Lodge said.

As far as Richet and Ochorowicz were concerned, the Cambridge sittings had been nothing more than a deliberate attempt to make Palladino look foolish, and consequently an attempt by their erstwhile British and American colleagues to make the two of them look foolish as well. They planned to continue studying her more objectively, without such "help."

The debacle had done nothing for the reputation of psychical research in the United States. Donations had dropped, and James was now paying Hodgson's salary out of his own pocket; so far it had cost him $300. "I fear the Eusapia business may prove a blow to our prosperity for a while," James told Sidgwick, "although Hodgson's withers are unwrung."

The lesson of the Palladino affair—the only lesson agreed upon by the investigators, anyway—was that the Italian medium possessed a genuine gift for causing trouble.

"THE PRESIDENCY OF the Society for Psychical Research resembles a mouse trap."

So began William James's farewell address, at the end of 1895, after two years of transatlantic presidency. He was gladly turning the mousetrap over to William Crookes who had accepted in advance. A good decade of solid scientific work had restored Crookes's sense of invincible self-confidence, and his attitude toward the importance of psychical research.

In recent years, Crookes had invented the radiometer, to measure particles in light; he would later invent the spintharoscope, which counted the alpha rays emitted by radium. He'd continued refining his Crookes tubes as well; they would prove useful not only in Thomson's experiments with electrons but also in the study of Roentgen rays (later to be known as X-rays).

Crookes's virtuosity with instruments kept him constantly in demand; most recently he'd collaborated with Lord Rayleigh, investigating a mysterious element in the atmosphere. Rayleigh's longtime interest in atmospheric studies had led to the isolation of the gaseous element that Rayleigh named argon, from the Greek word *argos*, "idle." Argon was a passive kind of gas, basically inert, which made it hard to tease out of the frothing chemical soup around it. But Crookes had done a spectrographic analysis that caused argon to glow like fire against its background; he'd labeled more than 200 lines of light associated with the gas. For his contributions to science, Queen Victoria would confer a knighthood on Crookes in 1897.

With his reputation for science and sanity more than restored, Crookes now chose to declare that he still believed in the supernatural. He stood by his earlier investigations of D. D. Home, he said, and he stood by his convictions. In accepting the presidency, he had only one new goal. He wanted to convince his fellow scientists to try for a little more humility, to let go of their "too hasty assumption that we know more about the universe than we can possibly do."

William James also spoke of the need for humility in his address—but in psychical research itself. While his colleagues had definitely made progress in establishing telepathy and crisis apparitions, they had yet to find a mechanism that would explain them, that *modus transferendi* that Sidgwick longed to discover.

Their real accomplishment, he thought, was to establish psychical research as a field whose questions merited answers. That success came mainly through the steady work of the British society and its building of theoretical connections between telepathy and apparitions; and, he thought, the work of Richard Hodgson and others, including himself, in doing a case analysis of Leonora Piper. Through that sharp focus on mediumistic powers, James thought they were learning lessons that might be applied in a much wider sense. "If you will let me use the language of the professional logic-shop, a universal proposition can be made untrue by a particular instance. If you wish to upset the law that all crows are black, you mustn't seek to show that no crows are; it is enough if you prove one single crow to be white."

In the case of admitting the supernatural, he continued, "My own white crow is Mrs. Piper. In the trances of this medium, I cannot resist the conviction that knowledge appears which she has never gained by the ordinary waking use of her eyes and ears and wits. What the source of this knowledge may be I know not, and have not the glimmer of an explanatory suggestion to make; but from admitting the fact of such evidence I can see no escape."

Given such accomplishments, James then pondered, why had their work been so steadfastly dismissed and belittled? As he saw it, the answer lay not with them but in the nature of nineteenth-century science, its reliance on fixed laws, and its "belief that the deeper order of Nature is mechanical exclusively."

A Harvard-educated scientist himself, William James believed in rules; he believed that the scientific worldview provided enormous benefits to humankind. From nineteenth-century science had come vaccinations, a new treatment for diabetes, pain-soothing anesthesia, the telephone, the telegraph, the phonograph, the newly designed internal-combustion engine, electricity in homes and businesses, and a future filled with the promise of more and better. The benefits of science were unquestionably great, James said, and "our gratitude for what is positive in her teachings must be correspondingly immense."

But to return to the theme of humility, even such a string of successes didn't mean that scientists held the universe and all its secrets in their hands. The real shortcoming of science in the dawn of the twentieth century, James said, was its rejection of all experiences and insights not generated by the priesthood of science itself.

In the short term, he thought, such arrogance would allow the research community to establish a new level of power and influence. But in the future, James wondered if those who were now awed by innovation and cowed by superiority would continue to be so malleable. It might be that those so summarily dismissed by scientific leaders might someday dismiss those leaders just as conclusively. "It is the intolerance of Science for such phenomena as we are studying, her peremptory denial either of their existence or of their significance except as proofs of man's absolute folly that has set Science so apart from the common sympathies of the race."

James warned that future generations of scientists might pay a price for the intolerance of his time, that respect and admiration for research could not continue indefinitely without some respect given in return. He wondered if scientists of the twentieth century would regret the lost opportunity to share in a societal discussion. Some day in the future, James concluded, a more enlightened society would mourn that determined blindness in "our own boasted Science, the omission that to their eyes will most tend to make it look perspectiveless and short."

9

THE UNEARTHLY
ARCHIVE

*A*S THE TWENTIETH CENTURY drew closer, gleaming with all the bright sheen of well-polished metal, Fred Myers found himself looking away, back toward his past and his youth. He recalled gentler times, softer days, and the vanished music of Annie Marshall's voice.

Myers turned fifty-three in 1896; in the looking glass, he saw a middle-aged man with a silvered beard and a level dark gaze. He was steadier now, more serious in his outlook, more determined to help resolve the philosophical contradictions of the world around him.

His wife tolerated his psychical obsessions, he knew, but she did not share them. Evie didn't pretend to be an intellectual or a seeker. Her life turned around their children, her photography, and cherished social functions. She depended, as she told him, on his superior intelligence and strength. Myers didn't tell her that sometimes he longed to be less of a mainstay. He didn't tell her either that what he missed most from his more impetuous days was another love, a woman with sad blue eyes and a smile like sunlight.

"Do not allude to all this in any letter," Myers wrote to James, because "my wife likes to see your letters." The previous year, when he'd come over for the Chicago meeting, Myers rode the train to Boston so that he could ask Mrs. Piper to try to find Annie Marshall, lost from his sight in the spirit world.

In another secretive letter, Myers confided in Oliver Lodge: "I do not say that facts *unknown to myself* were given[,] but facts unknown to Mrs. P were *recombined* in a manner & with an earnestness which in Hodgson and myself left little doubt—no doubt—that we were in the presence of an authentic utterance from a soul beyond the tomb."

Myers asked Richard Hodgson to continue working with Mrs. Piper in search of further proof. It would be easy enough to abandon this ridiculous, hopeless wish to find a dead woman. But Myers, as he told his friends, had a feeling that she was there, just beyond his reach.

Myers found Leonora Piper fascinating anyway. The hunt for Annie just added to the medium's attraction as an object of study. The mind of a medium, Myers would argue, offered a rare opportunity to explore the true range of human capabilities.

He compared Mrs. Piper's mental manifestations when in trance state—personalities including G.P. and Phinuit—to interlocking puzzle pieces that formed a picture of an intricate brain, operating on levels that ranged from waking awareness to "subliminal consciousness." He'd already published eight arguments in favor of the multifaceted mind, filling page after page in the *Proceedings of the Society for Psychical Research*. But Myers wanted to go beyond mere advocacy of his position; he wanted to explore the potential of such a mind. How might different aspects of the brain, different levels of its function, influence each other?

Myers pored over other writings on the subject, sifting them for support. He persuaded the SPR to publish Sigmund Freud's 1895 paper "Studies in Hysteria" because the Austrian discussed the idea of the subconscious mind. It was the first of Freud's research papers to appear in a British journal. Freud was just building his reputation as a pioneer in psychiatry at that moment; he would first use the term *psychoanalysis* in 1896, after ten years of private practice. He had studied hypnosis under Jean Charcot, in Paris, and would later refer to the hypnotic state as one of free association. When the SPR first published his study, Freud had yet to make an impact in London, although his provocative theories were gradually gaining attention. Myers particularly liked Freud's innovative ways of looking at mental processes. "The fact is," Freud wrote in "Studies of Hysteria," "that local diagnosis and electrical reactions lead nowhere in the study of hysteria," that a person had to be understood in terms of their life, their experience, their story, that dismissing them as a pathology was too shallow an approach.

Myers saw nothing shallow or simple about the human mind. He frequently compared the range of human consciousness to the light spectrum, in which visible light constitutes only one small part, while other regions—from the ultraviolet to the infrared—do their work in ways that we cannot see but may sense and respond to anyway. Using that image as an illustration, he offered his own theory proposing that ordinary consciousness, which he sometimes called "supraliminal," constituted only a small part of mental abilities, the part focused upon helping us function in our daily lives.

Beneath that "waking self" ran the other "streams of consciousness," currents of mental activity just as invisible as the heat-rich radiation of the infrared region of light and just as potent. Myers proposed that some part of the subliminal, or subconscious, mind was also purely about internal function—managing digestion, the thump of the heart, the whoosh of the lungs. Other "streams" of subliminal consciousness, though, might be used for external functions—processing social communication signals such as facial expression or subtle body gestures. At other levels, the subliminal mind might operate even more subtly, even telepathically. "I suggest, then," he wrote, "that the stream of consciousness in which we habitually live is not the only consciousness which exists in connection with our organism."

Perhaps, Myers added, these "other" consciousnesses could help explain what appeared to be mental aberrations: hypnotism, hysteria, inexplicable fears, crisis apparitions, and dreams. Perhaps even as people varied greatly in their waking intellect, so too did they differ in their subliminal intelligence. Perhaps there was even a balance to it; those brains most capable in the material world did less well in the subliminal realms—and vice versa.

That dichotomy, Myers proposed, probably held true for most people, including himself. The average person's brain focused entirely on handling life's obvious challenges, allowing little scope for developing telepathic skills, or any talents beyond the five senses. The few that did exercise their brains

in such areas—perhaps developing a skill in subliminal communication—
might find that it cost them in other areas, organized their mind in a way
that left them with less potential for an academic style of intelligence. Such
people might even appear peculiar in the everyday sense. They might be
prone to trances, subject to developing odd trance personalities. Those with
a strong subliminal life might, fairly or unfairly, be considered abnormal.
They might become mediums. Such a one might become a Leonora Piper.

THESE DAYS, William James found ideas like Myers's more interesting, or
at least more original, than those put forward by his fellow practitioners of
scientific psychology. His dissatisfaction spilled into a letter to Charles
William Eliot, president of Harvard, assessing the field of psychology as
stodgy and depressingly lacking in innovation, especially for such a young
science.

He cited the University of Wisconsin's James Jastrow as an example.
Jastrow, who had been so argumentative about psychical research, did some
good experimental work, James wrote, but he had "a narrowish intellect . . .
and uncomfortable peculiarities of character." James hadn't quarreled yet
with Edward Titchener at Cornell University, but the man was unoriginal
in his thinking and "although from Oxford, quite a barbarian in his scientific and literary manners and quarrelsome in the extreme." The best psychologist at Yale was shallow; the University of Chicago had one promising
psychologist, but he was too young to have done anything interesting.

James admired James McKeen Cattell of Columbia University for his efforts to pioneer human intelligence testing. But Cattell's closed-mindedness
on the subject of psychical research frustrated James, especially when the
Columbia professor publicly chastised James for his support of the field.
Recently, Cattell had compared the SPR's work to a swamp, incomprehensible through the murk of superstition.

Cattell heightened his aggressive stance after reading a copy of James's
farewell address as president of the SPR. In an essay in the *Psychological
Review,* which Cattell had founded, the New York scientist expressed his
dismay that a respected psychologist such as William James could be so
fooled by the shoddy evidence and inadequate experiments offered up by

his fellow psychical researchers. As Cattell put it, there could never be real evidence for the supernatural because evidence could not exist for a fantasy. Stung, James replied to the *Review* that his critic clearly hadn't bothered to read the work: "The concrete evidence for most of the 'psychic' phenomena under discussion is good enough to hang a man 20 times over."

James was angry enough to add that such prejudice seemed unfortunately characteristic of the whole profession. Cattell's objections were typical of those raised by traditional researchers: shallow on their face, and "shallow by further investigation."

In an 1896 address to the Philosophical Clubs of Yale and Brown Universities, James further challenged the claim that the world could—or should—be understood only by application of logic and material evidence.

By taking refuge in "snarling logicality" James said, by insisting that the only believable god was one who meekly appeared when asked to prove himself, a person might "cut himself off forever from his only opportunity to make the gods' acquaintance." In terms of material logic, James said, it appears easier to disbelieve what cannot be proved. But by doggedly persisting, by remaining open to belief—in gods, deities, higher powers, purpose—we allow for opportunity. We keep our minds open to what we do not know for sure, to what we have no idea how to prove. In this, James said, it may yet turn out that, as believers, "we are doing the universe the deepest service we can."

James's talk would become the title essay in one of his first, and most famous, collections of philosophical musings. In *The Will to Believe,* he carefully explored the ways that humans can choose to understand—and to live in—the world. He gave a simple comparison: "Science says things are; morality says some things are better than other things," and religion says that the best things are eternal, "an affirmation which obviously cannot yet be verified scientifically at all."

This matter of choosing to live upon a foundation that could never be verified appeared to set science at odds with faith. James suspected that many scientists dealt with that challenge in the simplest possible way—by denying religious precepts entirely, without always asking themselves which intellectual pitfall was the greater evil: "*Better risk loss of truth than chance of error*—that is your faith-vetoer's exact position. He is actively

playing his stake as much as the believer is; he is backing the field against
the religious hypothesis, just as the believer is backing the religious hypothesis against the field."

He argued that pursuit of truth, even when it might seem illogical by
the rules of science, was always worth the risk. For himself, James had
decided against "agnostic truth-seeking"; he couldn't suppress his belief
that to reach the stubborn mysteries of the universe, one had to be willing
to believe that the most unlikely paths might lead in that direction.

FOR ALMOST TEN YEARS, William Barrett had held tightly to his grudge
against the Society for Psychical Research. He refused to forgive either the
late Edmund Gurney's exposure of some of Barrett's favorite telepathy subjects or Sidgwick's insistence on striking some of his favorite experiments
from the SPR record.

He had remained a member of the organization, but a sulky and unco-
operative one. Fred Myers had further stoked Barrett's resentments. Myers
was forever writing notes demanding that Barrett better document his private psychical research. Barrett disliked everything about those notes—their
chastising tone, the way that Myers underlined his criticisms with an insulting slash of black ink, the fact that they were written by a philosophy
major. He might have abandoned the whole cause if he wasn't even more
irked with his colleagues in orthodox science.

"About the narrowest minded, most intolerant & least sympathetic
minds at present," Barrett wrote to Oliver Lodge, "are those whose eyes are
forever glued to the microscope of their own special branch of science."
Lodge agreed. He also saw an opportunity in the Irish physicist's evident
exasperation. He suggested to the Sidgwicks that if approached nicely, Barrett might be ready to work directly for the SPR again.

Henry Sidgwick had just the project in mind. He wanted someone to
investigate the never-ending claims about dowsing or "divining" rods and
their users. As the SPR had gained visibility, Sidgwick found himself field-
ing a stream of letters concerning dowsers and their supposedly eerie spirit
powers. Many of these letters originated in Cornwall, where the rods were
thought to be especially responsive to pixies. Sidgwick wrote to Barrett,

offering to personally finance an objective investigation of such claims. Barrett recognized this as Sidgwick's notion of an olive branch.

After he'd agreed to do it, even after he'd completed the investigation, Barrett confessed to personal doubts about the task. "Few subjects," as he admitted in the resulting 1897 report, "appear to be so unworthy of scientific notice."

Divining rods seemed carved equally from wood and from superstition. They'd been used for centuries to find water and more—mineral lodes, buried treasure, archaeological sites, even criminals. The Romans thought the rods were blessed by the god Mercury; the Germans had called them "wishing rods" and used them to tell fortunes. There had been early Christians who believed diviners were blessed by God. Others claimed the rods worked best when cut from a tree on either Good Friday or Saint John's Day. But Saint John's Day, June 24, corresponded roughly with the summer solstice, an ancient pagan time of celebration. Some believed that both the rods and their users were un-Christian, possessed by the devil.

Divining lore—sometimes called rhabdomancy (from the Greek *rhabdos,* "rod," and *manteia,* "divination")—declared that, handled correctly, the rod would move as truly as a magnetic needle pulled to the north. At the moment of discovery, the rod's operator (or rhabdomancer) would feel its power: sudden acceleration or retardation of the pulse, a sensation of heat or cold prickling along the skin. At the same time, the rod would bend, pointing toward what was sought.

The wood of choice was traditionally cut from the hazel tree, with its reddish bark, heart-shaped leaves, gold-colored hazelnuts, and mystic history. In Celtic lore, nine hazel trees guard the entrance to the Otherworld, and their nuts serve as containers for wisdom itself. A good divining rod was cut with a branched fork at one end, so that it could be held two-handed. Diviners held one hand curled around each prong of the fork, leaving the attached branch to extend downward like a flexible wand. But most people preferred a simpler term for a simpler version of the rod, a single stick known as a magic wand.

Despite his willingness to entertain the notion of telepathy, Barrett definitely did not believe in magic, and he considered rhabdomancy to be nonsense. In the case of water dowsing, in particular, he suspected that good diviners merely possessed a sharp eye for changes of vegetation and other observable signs of underlying wetness.

Barrett began his study by gathering records from the history of divining, finding to his satisfaction that most discoveries were easily discredited. But—as it turned out so often in psychical research—a handful of reports stubbornly held up, despite his best efforts to see and expose fraud.

There was, for instance, an 1889 case from the operators of a business in Ireland, the Waterford Bacon Factory. At that time, the factory managers had been working with geologists to drill a new well. Numerous holes had been sunk, one to a depth of 1,000 feet, and all had been dry. In frustration, despite the outrage of the consulting scientists, the managers decided to consult a dowser.

The man arrived on a sunny afternoon, forked rod in hand. He ambled around the site, his head cocked a little, as if he were listening, until suddenly the twig twisted so sharply that it snapped in two. Not only did the dowser insist water lay below, he provided a depth range between 80 and 100 feet. The Irish Geological Society had assigned a representative to observe this tomfoolery; he gloomily reported back that when the new borehole was sunk, as directed, water bubbled up. The resulting well yielded up to 5,000 gallons an hour.

The good examples, and they were rare, were all like that. There were reputable observers, witnesses who swore that the whole exercise had been a gamble, that they'd expected no more than a good show, right until the moment that the rod leapt to life in its owner's hands, like a thing possessed.

Barrett picked apart most divining claims without ceremony. He dismissed the reputed motion of the diving rod as an involuntary muscle spasm; the dowser's arms would get tired holding a stick aloft indefinitely. He still suspected most water discoveries involved knowledge of the area, visible evidence of water. Further, the record keeping was so poor, he complained, that it was impossible to compare the overall success of dowsers to that of geologists engaged in the same pursuit.

And yet he didn't deny that a few cases—such as the Irish well-drilling episode—could not be so simply explained away. He saw no evidence for magic there. But he thought a few operators might possess an unusual sensitivity, perhaps another example of one of those subliminal selves proposed by Myers. It might be, Barrett thought, that it was a kind of innate ability, comparable to migratory or homing instincts in other animals. The ability to somehow sense the presence of water, after all, would have been a powerful evolutionary advantage if it had occurred in ancient mankind.

As he had twenty years before, in his first paper on telepathy, Barrett concluded his report for the SPR with a recommendation. Here was a puzzle, he wrote, toward whose solution a scientific investigation might serve a useful purpose.

RICHARD HODGSON HAD also been thinking over innate abilities, trying to put into perspective the mind of a medium and what it could—and couldn't—accomplish. Surprising himself and his colleagues, Hodgson announced that he'd been mistaken in calling G.P. merely a trance personality of Leonora Piper. After sittings with 130 different visitors, he'd been persuaded of the impossible—that the personality in the room was indeed a spirit, proof that his friend lived on.

Out of that long line of visitors, only twenty or so were friends of the late George Pellew. The rest were strangers, brought in to muddle the picture. All were presented without a clue as to name or background. Yet G.P. had effortlessly sorted through this parade, greeting all his old friends by name except one, a girl who was now eighteen and had been only ten when he met her. She had changed, G.P. told her finally, adding rather rudely that he wondered if she still played the violin as badly as she had as a child. As Hodgson reported, not once in the years between 1892 and 1897 did "G.P." ever confuse a stranger for a friend of George Pellew—or vice versa.

Hodgson found telepathy an inadequate explanation; it could hardly be supposed that all of G.P.'s friends happened to be gifted telepathic agents, capable of sharing their thoughts with the medium. Sometimes G.P. talked accurately about friends not in attendance, some living miles away, making thought transference even more unlikely. Hodgson found support from other sittings as well. For instance, messages from people who had apparently died in mental anguish, such as suicides, were consistently confused, almost desperately so. If Mrs. Piper worked by telepathy to create a mental picture of a person lost to suicide, by reading the minds of friends and acquaintances, there was no reason that it would be garbled compared to all the other mind-reading. Time after time, though, messages from suicides remained muddled, miserable.

By contrast, there continued to be occasional sittings that rendered such breathtakingly clear and personal responses that even an observer given to doubt could not avoid that sense of a spirit in the room. In one such sitting, the parents of a little girl, Katherine (nicknamed Kakie), who had died a few weeks earlier at the age of five, came to visit. They did not identify themselves but brought with them a silver medal and string of buttons that the child had once played with.

A transcription of the sitting read as follows:

> Where is Papa? Want Papa. [The father takes from the table a silver medal and hands it to Mrs. Piper] I want this—want to bite it. [She used to do this.] . . . I want you to call Dodo [her name for her brother George]. Tell Dodo I am happy. [Puts hands to throat] No sore throat any more. [She had pain and distress of the throat and tongue] . . . Papa, want to go wide [ride] horsey [She pleaded this throughout her illness] Every day I go to see horsey. I like that horsey . . . Eleanor. I want Eleanor. [Her little sister. She called her much during her last illness.] I want my buttons. Where is Dinah? I want Dinah. [Dinah was an old rag doll, not with us]. I want Bagie [her name for her sister Margaret]. I want to go to Bagie . . . I want Bagie . . .

It was a shock to hear Hodgson, the longtime cynic, the tough-minded investigator, the most skeptical of the SPR inquirers, declare such sessions to be evidence of spirit communication. But once convinced, he did so with typical forthrightness in his second report on Leonora Piper, published in December 1897: "At the present time I cannot profess to have any doubt but that the chief 'communicators' to which I have referred in the foregoing pages are veritably the personages they claim to be, that they have survived the change we call death, and that they have directly communicated with us, whom we call living, through Mrs. Piper's organism."

Hodgson's paper illuminated beautifully—if unintentionally—the inherent difficulties of producing persuasive results in psychical research. To accept that G.P. was a spirit, one had to believe in immortality. Further, one needed to believe that the exchanges between this modest American medium and a self-proclaimed spirit proved the reality of life after death, trumped all other explanations.

Henry Sidgwick had longed for the day that his society showed the skeptics wrong, delivered up indisputable proof that the soul survived. But he could not convince himself that this was it; he'd spent too many years picking apart evidence, and he could see ways to pick apart Hodgson's report as well.

The G.P. sittings were remarkable, Sidgwick agreed, but they did not entirely exclude telepathy. True, it seemed unlikely that all G.P.'s friends were strong telepathic communicators, delivering mental information to Mrs. Piper. But it was not impossible, and therefore thought reading could account yet for recognition and revelations about the dead man's friends. Further, Sidgwick was troubled by the fact that while G.P. recognized others so readily, he retained so little knowledge of himself, at least of his former intellectual pursuits. Pellew had been an avid student of philosophy; the trance personality barely recognized the subject.

In one interview, a visitor asked G.P. about the American philosopher of science Chauncey Wright, who had been one of the earliest defenders of Darwinism and who had warned that theology should expect little support from the laws of science. Wright had written, for instance, that the idea that "the universe has a purpose" could be believed on grounds of faith only. It could never, he said, be "disclosed or supported" by scientific investigation. Until his death in 1875, Wright had worked in Cambridge, and his philosophical essays were widely read by northeastern intellectuals, Pellew among them.

Asked whether G.P.'s life after death shed light on Wright's views of natural laws, the session went as follows:

G.P.: Yes, law is thought.

Sitter: Do you now find that law is permanent?

G.P.: Cause is thought.

Sitter: That doesn't answer it.

G.P.: Ask it.

The sitter asked if G.P. agreed with Chauncey Wright and was first
    told "most certainly" and then told, "He knows nothing, his
    theory is ludicrous."

Surely, Sidgwick argued, the real spirit of George Pellew could have
handled simple questions about a philosopher whose work he knew well.
Did such hard-earned knowledge just leak away, sand trickling from broken glass, once a person died? Sidgwick found it hard to accept that the
mind might survive but only as an empty container, bare of the knowledge
that once filled it.

If Hodgson had hoped for better support from his colleagues, he did
not chastise them for their doubts. Instead, he set about answering the criticisms. Sidgwick, he said, had raised one of the more interesting and com-
plicating aspects of spirit communication, the difficulty of communicating
through a medium. It called to mind the challenge of the "ghost of clothes"
question, the way that one mind may alter information received from
another. As Hodgson pointed out, Mrs. Piper knew nothing of philosophy.
She was unlikely to understand it or relay its finer points with any grace.
Her ability was to receive these flickers of communication but she wasn't
necessarily a competent interpreter. "If Professor Sidgwick were compelled
to discourse philosophy through Mrs. Piper's organism, the result would
be a very different thing from his lectures at Cambridge," he emphasized.

As Hodgson considered the issue, he'd come to believe that some
things might be easier for spirits to communicate than others. Emotional
connections—with their pure, personal power—might survive fairly intact
through the translating mechanism of the medium. Intellect and sophisticated knowledge would be unlikely to fare so well, especially if the transla-
tor were uneducated, or if the medium lacked the language and training to
understand what was being said in the first place.

He reminded Sidgwick of all the obstacles that must be overcome for
any spirit communication, even of the most primitive type, to occur. If one
considered the difficulty of communication between two living people in

the same room—the way one person interprets or misinterprets another's thoughts during a conversation—how much more difficult to conduct that conversation with someone speaking from another dimension, using the awkward device of an entranced medium to relay messages? "The conditions of communication must be kept before the mind," Hodgson insisted, and expectations for fluency should be lowered as a result.

JAMES McKEEN CATTELL, the Columbia professor who had so vehemently disparaged James's SPR presidential address, read Hodgson's affirmation of spirit life and hated it from first paragraph to last.

The life-after-death insinuations in Hodgson's report struck him as simple spirit-mongering, and its conclusion—that even skeptics such as Richard Hodgson could be converted—infuriated him. Cattell didn't really care if Hodgson wanted to make a fool of himself. But given the author's reputation as a savvy investigator, Cattell did worry about the report's influence on the beliefs of other scholars. What if a reputable scientist were to conclude that since the previously far-from-gullible Hodgson had crossed the line to credulity, it had become an intellectually permissible, even a respectable ideological crossing? The prospect horrified Cattell.

He fired off an essay to *Science* magazine, titled "Mrs. Piper, The Medium" (in homage to Browning's cynical portrait of D. D. Home), to make sure that the real scientific point of view was understood. Cattell took aim not only at Hodgson's analysis but at what he considered the bigger target, William James's support of the SPR studies. Referring to James's earlier description of Mrs. Piper as "the white crow" that helped persuade him of supernatural realities, Cattell wrote: "The difficulty has been that proving innumerable mediums to be frauds does not disprove the possibility (though it greatly reduces the likelihood) of one medium being genuine. But here we have the 'white crow' selected by Professor James from all the piebald crows exhibited by the Society." Her credibility was due not to her own talents, Cattell continued, but to being endorsed by one of the country's premier psychologists.

Psychical research was clearly costing William James academic prestige and political capital among his fellow scientists. Privately, he confessed some regrets over it. Publicly, James responded as if he didn't care. He wrote back to *Science,* characterizing Cattell's position as a childishly simple argument that "mediums are scientific outlaws and their defendants are quasi-insane," going on to suggest that the magazine's readers might prefer more intelligent, sophisticated criticisms. For the discriminating reader, James recommended Sidgwick's dissection of the G.P. case, which could be found in *Proceedings of the Society for Psychical Research.*

Continuing that acerbic exchange, Cattell dismissed the opinions of nonscientists, and especially those belonging to the SPR, which he said was doing active harm, encouraging people to cling to the mysticism of the past. The role of science, he said, was not to pander to superstition but to help eliminate it. And when a leading psychologist such as William James failed to live up to that role, he could be held personally responsible for holding back progress itself.

"I believe that the Society for Psychical Research is doing much to injure psychology," Cattell concluded. "The authority of Professor James is such that he involves other students of psychology in his opinions unless they protest. We all acknowledge his leadership, but we cannot follow him into the quagmires."

James could handle vitriol like this with ease. He replied mockingly that he enjoyed Cattell's "amiable persiflage" and feeble attempts at insult. It bothered him, though, that he could not persuade his peers to see the value in psychical research. Further, both Hodgson's report and the SPR response conformed to scientific principles. Hodgson had offered a theory and the supporting evidence for it. His SPR colleagues had reviewed it, criticized it, and demanded more substantial evidence.

James himself tended to side with Sidgwick in terms of the report's shortcomings. He didn't deny that G.P. provided some startling moments, but the personality also showed the same "vacancy, triviality and incoherence of mind" that so often plagued the spirit messages from Mrs. Piper and, indeed, those from all mediums. Hodgson's attempt to excuse such meanderings by extreme difficulty of communication struck James as inadequate, as he made clear, again writing in the SPR journal, "Mr. Hodgson has to resort to the theory that, although the communicants probably are spirits, they are in a semi-comatose or sleeping state, while communicating, and only half-aware of what is going on and Mrs. Piper's subconscious is then forced to fill in the gaps of whatever they say." This seemed at best an imperfect cover story. Even worse, the explanation discounted the best sittings in an effort to excuse the poorer ones. What about those apparently pitch-perfect days? Did the spirits suddenly wake up? Did Mrs. Piper's hearing improve? Could she briefly understand the ghostly communicators better?

If Mrs. Piper didn't cheat—and no evidence yet existed that she did— then it was still unclear to James how she accessed the information revealed in her trances. He continued to believe that she possessed some exceptional power; he continued to have no idea exactly what that power might be.

"If I may be allowed a personal expression of opinion at the end of this notice," James said, "I would say that the Piper phenomena are the most absolutely baffling thing I know."

DESPITE THE DOUBTS of his colleagues, mostly thanks to his reputation as "an expert in the art of unveiling fraud," as the *Saturday Review* put it, Hodgson's latest Piper report received exactly what Columbia's Cattell had feared, serious attention.

At the *Review,* famously hostile to psychical research, the editors wrote to acknowledge that Hodgson's account of G.P. provided strong evidence in favor of survival after death. Still, the *Review* emphasized, it was unclear exactly *what* survived, whether it was a soul, a spirit, or merely some sort of imprint of a personality. As the editorial noted, "So far as we can see, all that is proved is that some record of the life on earth is laid up in some unearthly archives, and that under some circumstances, this record is accessible to the minds of the living."

G.P.'s knowledge of his life on Earth, especially his previous relationships, seemed remarkable. But the "spirit" continually failed to provide any real detail about life after death. His descriptions, "while free from the nauseous sentimentality mingled with Swedenborg which forms the bulk of so-called spirit communications," were either vague or comfortably Christian, adding nothing new to the knowledge of immortality.

In conclusion, the magazine raised a point of elegant metaphysics: "The question is not whether something survives death, but whether that is a living something; whether it grows? Time may give us an answer to the question; but it has not been given yet."

That unearthly archive—or at least the possibility of something like it—was an idea that William James had considered, hoping that it might solve some of the troubling questions in psychical research.

It was just the first glimmer of a thought, really, but James wondered whether the energy generated in our lives—with all their passion and grief, laughter and argument—did more than fall to dust. Perhaps life's energy burned an impression, or memory, a cosmic record of sorts that lingered after the person himself had vanished.

Perhaps the very objects that we handle could sometimes be energy repositories, absorb some of life's stray heat, radiate it back out. If so, that might explain the improbable art of psychometry, the occasional flash of insight that a good psychic seemed to get from holding a piece of jewelry or an article of clothing. It might even explain haunted houses, those curious impressions of spirits that tended to repeat over decades, even centuries.

And perhaps—as the editors of the *Review* posited—that added up to a different explanation of immortality. Perhaps there was no real life after death, just the occasional echo of what was, sounding briefly in the night and fading away.

Pursuing that set of ideas, James proposed that most of us never hear the echoes at all. We live sheltered, born with mental buffers—or dikes, as he called them—to protect against such intrusions, to keep life from being too impossibly strange. But sometimes—as with a crisis apparition—that last blast of desperate energy overcomes those barriers so that just for a moment we hear our dying mother's voice, see the face of a lost friend.

James had recently evaluated just such a case, the story of Bertha Huse and Nellie Titus, which seemed to capture those possibilities. It was all there, the young woman's unseen fall into a lake, the body trapped out of view, the dream image of the tragic accident. There had been no conversation in the dream, no purposeful ghost, merely an intense image of the girl's last moments. Mrs. Titus reported a history of such dreams, flashes of insight caught in the quiet night. Perhaps in her undefended sleep, she was unusually open to those energy surges created by a final moment, allowing her to receive what seemed to be a message from the dead.

The same explanation might also serve for the medium James knew best, Leonora Piper. Perhaps she was even less well defended from such signals, more prone to picking them up on a frequent basis. Both women might belong to a small group of people born without adequate mental barriers to that cosmic record, so that "fitful influences leak in, showing the otherwise unverifiable common connection."

BAFFLING PSYCHIC PHENOMENA were something Charles Richet understood well, too well. In the world of traditional science, he knew what he was doing and that his work was good. He was exploring the immune system, probing the mechanism of fever. He was testing treatments for the great killer tuberculosis. In his spare time, he continued to design flying machines—motorized gliders—and test them himself, looping like a crazed moth around his family's summer home in southern France.

In the world of psychical research, however, Richet felt curiously on the defensive. From Boston, Richard Hodgson had made him look a fool over Eusapia Palladino. And now from England, Frank Podmore, one of the collaborators on the SPR's respected *Phantasms of the Living,* had taken an aggressive stand against telekinesis—supposedly one of Eusapia's talents—basing that position on a recent study of reputed poltergeists.

Podmore had picked apart eleven reported poltergeist cases in England, concluding that the flying objects and crashing furniture could be attributed simply to girls seeking attention. Often, Podmore sneered, the talents of psychical researchers were not required. Police officers sufficed.

He provided an example from a British country village, where an agitated couple declared that after their niece moved into their home, doors suddenly rattled in the night, windows crackled, furniture trembled out of place. Frightened, the aunt and uncle called the police. A constable hiding outside a window was able to see the girl tapping on the glass when she thought no one was looking. Another saw her kick off her boot, scream, and claim that the spirits had taken it.

Almost every poltergeist case in Podmore's inventory involved a female in apparent need of attention. As he reported in the *Proceedings of the Society for Psychical Research,* there was no reason to invoke the mysterious powers of telekinesis: "Naughty little girls" provided a perfectly good answer, and no one needed to invent a new kind of spiritual energy to explain them.

Yet that was precisely the power that Richet hoped he could demonstrate in Eusapia Palladino. Goaded, he began a new series of experiments, joined by Ochorowicz, a new colleague, a cautious Swiss psychologist named Theodore Flournoy, and a trio of Italian scientists. They were aided by Eusapia herself, who also felt humiliated by the Cambridge experiments and was willing to submit to extreme measures if they would remove the shadows from her name.

In one brutal series of tests, devised by physiologists from the University of Naples, the experimenters bound Eusapia's hands and arms with cords. They tied the cords onto iron rings in the floor and dripped lead seals onto the knots. As Richet reported, even after being bound for hours, she was able to summon those odd ectoplasmic hands—"some frail and diaphanous, some thick and strong"—all of which dissolved like mist when touched.

In a letter to Lodge in the fall of 1898, Richet repeated his conviction that Eusapia possessed some kind of power. It was erratic, yes; uncontrollable, even by her, yes; complicated by her devious nature, yes; but real. Lodge replied apologetically that it sounded intriguing, but he was at the moment overwhelmed by his work in wireless communication. But Fred Myers decided that he was weary of his overcautious colleagues.

To Richet's pleasure, Myers agreed to sail to Paris and meet once more with the controversial medium. Enthused, Richet again enlisted the help of Theodore Flournoy. A kindly man and a thoughtful scientist, Flournoy had become more deeply involved in psychical research after striking up a friendship with William James at the experimental psychology meetings. It was Flournoy who wrote to James, describing the Paris sittings. Myers's presence, he informed his friend, "gave much zest to the first séance because Eusapia was obviously bent on convincing him, after the unfortunate séances in Cambridge two years ago."

Richet kept the lights bright, using both an unscreened lamp and a blazing fire to illuminate the room. Flournoy could see "every finger of Eusapia; every feature; every detail of her dress." The séance was quiet throughout. Again the curtains blew with that odd, invisible wind. Again cloud shapes formed in the room, touching, brushing by, and dissolving like mist around them. By the end of the sittings, "Mr. Myers declared himself convinced," Flournoy wrote, "and I don't hesitate to agree with him."

Myers returned to England fizzing with enthusiasm, eager to tell the story of the striking sittings in the *Proceedings of the Society for Psychical*

Research. Unfortunately, at least from his perspective, Richard Hodgson currently served as editor of the organization's journal. Hodgson flatly rejected Myers's article. In fact, Hodgson told Myers smugly, he'd commissioned an article evaluating mediums of the day; it was scheduled to list Eusapia "amongst the ranks of tricksters."

Myers stormed over to visit Henry Sidgwick, demanding that he overrule Hodgson. Typically—and Myers considered this an irritating trait—Sidgwick instead sought a compromise. He told Hodgson to drop his listing of fraudulent mediums. Such a list, Sidgwick thought, was unnecessarily combative. But he also rejected any article supporting Eusapia Palladino. She might impress Myers and Richet and Flournoy, but Sidgwick thought them too easily won over: "I cannot see any reason for departing from our deliberate decision to have nothing further to do with any medium whom we might find guilty of intentional and systematic fraud."

As Sidgwick reminded Myers, the SPR had yet to accomplish its most basic goal, convincing the scientific community to consider telepathy with a little respect. So far their most obvious successes had been outside the halls of academia, in the more welcoming walks of popular culture.

The effects of psychical research were so visible, across so many venues, that James wondered if they had gained a reputation as the foremost experts on spiritual matters. "This seems to be rather a grave moment for all of us," James wrote to Lodge. "We are changing places with a set of beings, the 'regular' spiritualists, whom we have hitherto treated with a species of contempt that must have been not only galling, but asinine and conceited, in their eyes." Stories of crisis apparitions regularly appeared in the daily newspapers; special editions dedicated to such stories rapidly sold out. Equally impressive were the creations of fiction writers—ghosts, demons, creatures of the nights—stalking the pages of magazines and books.

Some wrote to terrify, as did Ireland's Bram Stoker in his 1897 story of the evil undead, *Dracula.* Others spun satire. Oscar Wilde, who like Stoker was Dublin-born, had some years earlier published "The Canterville Ghost," a short story in which an American family discovers that its rented British mansion is haunted. The realization occurs after several days of trying to remove bloodstains from the library floor, only to have the horrid spots continually reappear.

After three days of scrubbing, as Wilde cheerfully wrote, "Mr. Otis began to suspect that he had been too dogmatic in his denial of the existence of ghosts. Mrs. Otis expressed her intention of joining the Psychical Society and Washington [their son] prepared a long letter to Messrs. Myers and Podmore on the subject of the Permanence of Sanguineous Stains when connected with Crime."

Novelist Henry James Jr. also liked to spin ghostly tales, not surprisingly, considering who his father had been and who his brother was. Henry had published his first thriller in 1868, some twenty years earlier. In that creepy little tale, "The Romance of Certain Old Clothes," the ghost of a man's first wife kills the second wife, who happens to also be her scheming sister. More memorable, at least for Henry's brother William and the other SPR members, was "Sir Edmund Orme," which had been published as a magazine serial in 1892. The ghost of the title recalled the late Edmund Gurney, who had relatives living in London's Orme Court. The Edmund of James's rather vengeful tale had committed suicide when the woman he loved proved unfaithful. The story was set in Brighton, where Gurney had died.

Henry James confessed that he got the idea for his most famous ghost story while visiting Henry Sidgwick's cousin, Edward White Benson, who held the exalted position of archbishop of Canterbury but who was also a self-proclaimed ghost story addict. The archbishop held "ghost evenings" in his library, for friends to meet and tell tales, fueled by a good fire and plenty of alcoholic spirits. Benson's ghost evenings spawned many a literary venture. (His nephew, the famously satiric novelist E. F. Benson, was hailed for authoring some of the scariest stories of his time.) The story that Henry James began drafting after an evening of spectral tales at the Bensons was published in 1898 and was titled "The Turn of the Screw."

James started the story at a "ghost evening," narrated by a guest at a house party, one of a group enjoying an evening of spooky stories told around a fire. Most creatures of the night evoked in such stories are presented as if real. In James's artful hands, the ghosts were, instead, hauntingly ambiguous, evocative of all the unknowns that troubled his brother William.

In "The Turn of the Screw," a young governess secures a lucrative job caring for two children whose father often travels on business. Slowly, she perceives that the house where they live is haunted. The ghosts are silent, shadowy, but she comes to believe that they have come to carry away the children. She tries desperately to protect the young boy and girl. In the end, she fails. One child is hunted down by a vengeful spirit and dies in the arms of the governess.

Or so it might seem. But the interpreter of these threatening ghosts is the governess herself. The reader becomes aware that the spirits may exist only in her mind, and that the alternate story is of two children unfortunate enough to be put in the care of a psychotic young woman. Did she frighten her young charges to the point that one of them suffered heart failure—and died in her arms?

"Henry James has written a forceful story of country-home life," Myers wrote to Lodge that fall of 1898, in an ambiguous description of his own. Myers had no problem accepting the idea of ghosts. Such images permeated the SPR's records of crisis apparitions. The problem, the difference, was that the specters in James's story—if such they were—seemed imbued with a goal. "True ghost stories," Myers said, tended to be brief visions or sensations that flickered and vanished. They were startling, perhaps, but never really purposeful.

"Instead of describing a 'ghost' as a dead person permitted to communicate with the living, let us define it as a manifestation of persistent energy," Myers said. There was nothing in science to show that energy had a conscious purpose. The evil Count Dracula, the vengeful ghosts imagined by Henry James—these fictional manifestations bore little if any resemblance to what the SPR investigators had so far glimpsed.

IN 1898, WILLIAM CROOKES was elected president of the British Association for the Advancement of Science, another milestone in his return to mainstream science. Yet, he chose to give his presidential address, that October, on a note of blazing defiance.

In a speech in Bristol, to a materialist audience, Crookes deliberately returned to the subject of his favorite medium. Almost thirty years after he had published his first controversial account, D. D. Home's powers seemed to the scientist as compelling as ever. In retrospect, Crookes believed that Home had offered the first real demonstration of telekinesis and therefore confounded the scientific community. The late medium had proved, Crookes said, "that outside our scientific knowledge there exists a Force exercised by intelligence differing from the ordinary intelligence common to mortals."

Crookes was reclaiming ground from which he had long since retreated, after his unfortunate experience with London street mediums. Still active and productive as a physicist, he had that very year completed analysis of the fixation of atmospheric nitrogen; over the next few years, he would successfully undertake the separation of uranium isotopes and measurement of the radioactive decay process. His accomplishments had returned him to the inner priesthood, as Oliver Lodge had once described it. So, "perhaps among my audience some may feel curious as to whether I shall speak out or be silent," Crookes said.

"I elect to speak. . . . To ignore the subject would be an act of cowardice— an act of cowardice I feel no temptation to commit." Crookes wanted his fellow SPR members—and the greater scientific community as well—to clearly to understand that he still believed in supernatural powers and in his own experiments demonstrating them. "I have nothing to retract. I adhere to my already published statements," he said, and he found the more recent psychical research done by others equally convincing.

He believed in telekinesis; he believed in telepathy; he believed in the possibility that the dead might return. "Indeed, I might add much thereto."

Meanwhile, Hodgson's report continued to fulfill Cattell's fear that it would have the power to convert the undecided. Even worse, one of the more prominent converts came from Cattell's own institution.

JAMES HERVEY HYSLOP was a professor of philosophy at Columbia University, a slight man with a neat brown beard, chilly gray eyes, and the faint pallor of ill health, which had dogged him since childhood. The look of fragility was deceptive; he possessed the combative temperament of a pit bull terrier.

Born in 1850, Hyslop came from an Ohio farm family. He grew up in the tiny community of Xenia, a swatch of fiercely tended fields surrounded by forest. His childhood had been one of farm labor—from caring for horses to breaking away corn stalks after a winter frost—and ultraconservative Christianity.

His parents belonged to a fundamentalist Presbyterian church and followed its teachings to the letter. The children were required to study the Bible daily—although during the week they could also read certain newspapers and books. On Sunday, the whole family spent six hours attending sermons and memorizing Psalms. "We were not allowed to play at games, swing or whistle, ride or walk for pleasure, pluck fruit from trees, black our shoes or read any secular literature," Hyslop recalled. He'd followed those teachings faithfully as a child, but as a university student majoring in philosophy, Hyslop became convinced that his father's faith was at odds with reality. The son still accepted the notion of a deity. He could admit the "force of the argument for the existence of God or some intelligence at the foundation of things." It was the teaching of Christianity that now seemed to him preposterous—the impossibly simple explanation of creation, the egocentric notion of a chosen people, even the arguments for the divinity of Christ, seemed to Hyslop "fatally weak."

Determined not to be a hypocrite, he'd told his parents of his new perspective, proving to his farmer father that, as suspected, a university education led to godlessness. In the following years, Hyslop's father alternated between ignoring his son and bombarding him with warnings of damnation. Even after Hyslop received his Ph.D. in philosophy from Johns Hopkins, even after he was hired in 1889 as a professor of ethics and logic at Columbia, he knew full well that in his father's eyes he was a failure.

Hyslop fretted that he would never be able to repair the relationship, a loss made even more painful in 1896 when his father died of throat cancer. His father's death left Hyslop contemplating the rigidity of his opinions. Not about Christianity—nothing changed his mind about that—but about immortality. He began to wonder about survival after death, whether his father lived on in some form, whether he could reach him yet.

In early 1898, after reading Richard Hodgson's endorsement of Mrs. Piper, Hyslop realized that he'd found the medium through whom he could pose his questions. He wrote to Hodgson asking for a series of sittings designed to challenge Mrs. Piper's vaunted talents—and perhaps to resolve his personal dilemma. He proposed to make the challenge as difficult as he could. If it was too easy, it would convince no one, including himself.

As they arranged it, Hyslop not only attended the sittings anonymously, he wore a black mask over his face. He came masked even though he routinely waited outside a window until Mrs. Piper was in a full trance and Hodgson could gesture him into the room. Hodgson added another layer of protection to protect Hyslop's anonymity, a code name. Hodgson would refer to Hyslop only as "four times friend," since he had requested four sittings.

It was at the second sitting that Mrs. Piper told him that a spirit was newly arrived in the room, and that the visitor's name was Robert Hyslop. As Hyslop told Hodgson afterward, he didn't think four sittings were going to be enough.

THE SPR'S HOPES for convincing scientists that telepathy should become part of standard research had sustained yet another blow—and led William James into yet another public quarrel with one of his fellow psychologists.

The argument began after Sidgwick presented some new telepathy work—including a tidy set of experiments with playing cards, done by Oliver Lodge—at the summer's experimental psychology meeting in Munich. His presentation almost immediately provoked an article in *Science,* suggesting that the SPR telepathy subjects cheated their way to success, possibly by simply whispering the correct cards to each other. Although the author later admitted that he'd not really proven that case, the article was widely praised by scientists and hailed in a letter from psychologist Edward Titchener of Cornell University, which declared, "No scientifically-minded psychologist believes in telepathy."

Once again, James took up the cudgels for his friends and his beliefs. He wrote to remind the readers of *Science*—and Titchener—that the original author had backed down from his first assurance that fraud alone could explain the SPR results. "Even in anti-telepathic Science accuracy of representation is required, and I am pleading not for telepathy but only for accuracy," James said, expressing his regrets that Titchener was unable to meet that basic standard.

Insulted, Titchener replied that he *had* been accurate. Perhaps the author had backtracked a little, but at least he was a good scientist, as opposed to the slipshod variety found in the psychical research community. Further, the basic point that "ordinary channels of sense," such as hearing a whisper, could account for so-called telepathic results was by far the preferred explanation.

Titchener and James had been leaders in American psychology for years. Both had studied under the great German experimentalist William

Wundt; both had persuaded their universities to establish their first psychology laboratories. It was true that Titchener's idea of psychology looked nothing like James's. He was a founder of structuralist theory—that the mind was composed of structures, such as thought and emotion, just as a water molecule was composed of structures, such as hydrogen and oxygen. He saw no place in the mind for a telepathy structure or a spiritual communication center.

But out of respect for their long-standing relationship, Titchener also wrote to James directly, trying to explain his viewpoint, providing an eloquent defense of his own position and of the stance taken by traditional psychologists. "I think that there is a great deal in your general position," Titchener began. "That is, I think that these topics have been boycotted, and should not be so." Still, he accused James and his SPR colleagues of constantly claiming persecution as a means of countering criticisms: "A minority is not always or necessarily in the right," Titchener pointed out. "And, together with many others, I rather resent the airs of martyrdom that psychical research puts on.

"You are perfectly free to work: you have a lot of big names on your side to back you up; your society is very flourishing. Suppose that Sidgwick and Balfour and the rest had done as much for psychophysics as they have for psychical research! Then there would be English laboratories worth the name." (Established earlier in the nineteenth century by German physicist-philosopher Gustav Theodor Fechner, psychophysics was the study of correlations between psychological sensations and the physical stimuli that trigger them.)

It was true that Titchener didn't plan to read SPR studies or conduct his own investigations into psychical phenomena. He would certainly never visit a medium, even Mrs. Piper, due to his "personal repugnance." If by fairness, James meant that Titchener needed to take a serious look at psychical research, he was afraid that they would never agree. "But I am as keen for fair play as anybody—meaning thereby that you have your right to fight for your side, and that I have an equal right to fight for mine."

AS IF TO ANSWER the worrying impression that only one decent medium existed to be studied, in the pages of the spiritualist journal *Light* appeared notice of a new trance medium, the twenty-nine-year-old wife of a London merchant. At first read, she seemed nothing special. Her spirit guide was reputedly a child, her daughter Nelly, lost years before, who spoke in the soft lisp of a toddler.

With Nelly came Eusapia-like effects—blowing curtains, flickering lights, the occasional levitations of furniture. The SPR would have ignored her except for one fact: Rosina Thompson didn't charge for the sittings, and they had agreed to give fair hearing to nonprofessional mediums.

Fred Myers decided to pay her a visit. And then another, and then another; he would eventually have 150 sittings and persuade the medium to abandon her physical productions and concentrate only on automatic writing. Once he accomplished that, to his surprise and pleasure, Mrs. Thompson produced the kind of results previously only seen with Leonora Piper. Myers wrote them up, in detail, for the SPR journal:

> The professor had come from Holland with a bundle containing a piece of clothing from a dead friend, a young man who after one unsuccessful suicide attempt—slashing his own throat but recovering—had shot himself to death the previous year.
>
> After Mrs. Thompson had slipped into a trance, he handed her the parcel. He had given neither his own name nor the name of the parcel's owner. As her fingers closed around it, Nelly's little-girl voice suddenly spoke:
>
> "I am frightened. I feel as if I want to run away."
>
> She set the parcel down and pointed at it.
>
> "This is a much younger gentleman. Very studified, fond of study. . . . He's not a rich gentleman. If he had lived longer he would have had more." He was worried about money, depressed, and headachy.
>
> All this was true, according to the professor from Holland. But it wasn't enough: "You have not told me the principal thing about this man."
>
> "The principal thing is his sudden death. . . . It frightens me. Everybody was frightened."
>
> She described the dead man, that he loved the outdoors, liked to hunt, and wore a round hat with a cord on it. All true again.

And then:

"I can't see any blood about this gentleman, but a horrible sore place: somebody wiped it all up. It looks black."

She was talking about the bullet hole. She described the cloth that had been put over his head when he was found dead. But it was the throat slashing that the spirit guide stayed fixed on.

"When any people want to kill themselves, he goes behind them. He stops them from cutting their throats. He says, 'Don't do that: you will wake up and find yourselves in another world haunted with the facts, and that's a greater punishment.'"

When Mrs. Thompson woke up, she complained bitterly of the taste of chloroform in her mouth. Myers's friend told him later that the chemical had been used in the treatment of the young man's slashed neck.

"My first sittings with Mrs. Thompson were in no way remarkable," Myers wrote to James in the fall of 1899. "There was little intimacy in the communications and Mrs. Thompson, as usual, came to herself with no recollection of the experience of the trance-state.

"But one day little Nelly announced the approach of a spirit 'almost as bright as God'—brighter & higher, at any rate, than any spirit whom she had thus far seen. That spirit with great difficulty descended into possession of the sensitive's organism—& spoke words which left no doubt of her identity."

Myers would not repeat the words—they were too private and too precious. He would not write the name of the spirit, although he knew that James would guess her identity. "May I not feel that this adoration has received its sanction, & that I am veritably in relation with a spirit who can hear & answer my prayer?"

He could almost hear Annie Marshall calling him. He found mostly joy in that, and a little fear as well. That shining spirit wanted him closer, it seemed, very much closer. Mrs. Thompson had written it down carefully, a promise that Myers would be reunited with his long-dead Annie—and soon—just on the other side of the dawning twentieth century.

10

ℳ

A PROPHECY
OF DEATH

*T*HE NEW CENTURY came in like sounding brass—a roar in the blood, a clatter in the ears, a triumphant drumbeat of progress. With the calendar turn to the 1900s, overseas phone calls arrived, along with double-sided phonograph records and Kodak's everyman camera, the one-dollar Brownie. German physicists introduced the idea of quantum theory; Freud published his revolutionary book *The Interpretation of Dreams;* the Zeppelin airship sailed through its first graceful test flight. And people hungered for news of more; in New York City alone, twenty-nine news-papers were hawked on street corners daily.

The twentieth-century personality was bright, loud, exhilarating, and, to fifty-eight-year-old William James, exhausting. He'd long fretted about his health, and now he felt depressingly old, a fragile man in robust times. Following a hiking vacation in the Adirondacks, he'd developed symptoms of heart disease. He could walk only a few feet before pain seared through his chest. On the recommendation of doctors—and given a new sabbatical leave by Harvard—James decided to seek medical treatment in Europe. Accompanied by his wife and daughter, he sailed from Boston in June

1900. Despite treatment at a renowned German spa, followed by bed rest at his brother's home in England, James could not seem to recover.

Fred Myers also was stubbornly ill. He'd emerged from a nasty bout with influenza only to baffle his physician with a persistent lethargy. Myers didn't have time to be sick, he told his doctors. He was writing a book on the subliminal self; one he hoped would forge a link between psychical and traditional research. He wanted to get back to it, if he could just find strength to put pen to paper.

Charles Richet, ever a generous friend, invited both the Jameses and the Myerses to make use of his chateau at Carquerainne, where he thought both men might benefit from the gentle climate of the French Riviera. He assured them that sunshine and sea breezes had been known to cure the most troublesome illness. The patients could recuperate together, Richet pointed out, and Myers could return to his writing as he grew stronger.

Myers felt a surge of enthusiasm at the prospect. While in France, he hoped to invite Rosina Thompson to conduct a few sittings. He liked the idea of getting James's opinion of this young medium. It might give him some perspective on her warning that death drew near. Myers sometimes thought that he could hear it closing in, the soft beat of wings, the approach of the angel of death, stirring the air behind him.

THE FRUSTRATING, fantastical Eusapia Palladino had risen from the ashes of her experiences with the SPR, thanks to Richet's continued championship. Again, she held court as the dominant medium on the European continent.

More than ever, she presented as an extraordinary specimen—uninhibited, tempestuous, erotic—a vision far removed from the sedate ways of the academic corridor, the neatly controlled setting of the laboratory.

Not only did Eusapia come out of trances charged with sexual energy, she sometimes seemed to shudder with pleasure while entranced. She claimed that, on occasion, the spirits brought her an invisible lover. A sly smile played across her face as she described, rather graphically, their encounters. She seemed to make the very air sparkle—and not just with figurative erotic energy. During one séance in Genoa, lights glittered overhead like dancing fireflies. One light settled on the palm of an observer, a German engineer; it was cool on his skin, he said, glinted briefly, and vanished even as he closed his hand about it.

The engineer—like physiologists, psychologists, and others from conventional academia—attended Eusapia's performances because Charles Richet had made curiosity about her permissible. Richet lent legitimacy to the Italian medium—much as William James's reputation had given Leonora Piper special status. His colleagues might deplore his interest in the supernatural, but geniuses were allowed their peculiarities, and Richet was a brilliant researcher.

His ongoing studies of the immune system were a case in point. The innovative French scientist would eventually find it necessary to invent a word for the allergic reactions that he had begun to study. He called the response "anaphylaxis," from the Greek words *ana,* "the opposite of," and *phylaxis,* "protection," describing a state in which an organism becomes oversensitized.

For instance, if a person were exposed to a particular poison—say, a bee sting or jellyfish tentacles—most would react the same way with each exposure. A sting would hurt, yes, but the victim would experience the same kind of swelling and pain with the first sting and with the tenth. A few individuals would become more tolerant, developing a kind of immunity to reaction. "The most remarkable case of this tolerance is to be seen when opium or morphine are used. People who take morphine injections need stronger and stronger doses for the morphine to take effect," Richet would explain. Others would become more sensitive with each exposure, so that if they were bitten or stung again—or gave themselves repeated injections—their body would overcompensate, even to the point of a lethal reaction.

Richet's multiyear inquiries into those varied responses—done partly by exposing dogs to repeated injections of jellyfish toxins—would open the way to the medical profession's understanding of anaphylactic shock. The work would also win him the Nobel Prize for Medicine in 1913 and a reputation as a world-class scientist long before that. Many of his colleagues wished that a scientist of Richet's caliber would abandon the peculiarities of psychical research. But his reputation made him—and his protégé—difficult to ignore.

.   .   .

MEDIUMS WERE PECULIAR creatures; there was no denying it about even the best of them. How could they not be? They spent hours of their time surrounded by people desperate to talk with the dead. They fell into trances reputedly inhabited by ghosts. They agreed to be hogtied by investigating scientists. Skeptics mocked them; journalists parodied them; former friends feared them. One had to wonder why anyone would choose to become a medium.

The sad and strange story of the Fox sisters was a case in point. Neither had become wealthy by pursuing such a career. Both had died paupers' deaths in the early 1890s—Kate at the age of fifty-two, Maggie at the age of fifty-five. Kate's body, reeking of old dirt and cheap gin, had been found on a sidewalk. Maggie had died in a tenement house in lower Manhattan, virtually alone.

Many in the spiritualist community had never forgiven the Fox sisters for their betrayal. Isaac Funk, of Funk & Wagnall's publishing house, expressed the widely held opinion that Margaret, especially, had betrayed the faith to feed her bad habits: "So low had this unfortunate woman sunk that for five dollars she would have denied her mother, sworn to anything," he wrote. But there were still those who believed that the Fox sisters had once been gifted, had been betrayed themselves by all those who used them for financial gain and promotional purposes. More than ten years after the Fox sisters died, schoolchildren playing in the abandoned cellar of the old Fox "spook house" found the complete skeleton of a man hidden behind a crumbling wall; apparently that of the murdered peddler they had first claimed to hear rapping. "Repeated [earlier] excavations failed to locate the body and thus give proof positive of their story," reported the *Boston Journal,* calling the discovery a reminder that not all about the Fox sisters had been false.

And a neighbor who stayed with Maggie Kane during the last week of her illness, in 1893, later told a curious story. The dying medium had been almost unable to move, crippled by rheumatism and weakened by fever. She mumbled constantly, asking questions of some unseen spirit in her rasping voice. As she spoke, knockings often sounded in the room, in the wall, the floors, the ceiling. There was no place to hide a rapping device.

The tenement room had no window, no closet, just a dresser, a table and chair, and a narrow cot with a ragged mattress. Upon her cot, the medium "was as incapable of cracking her toe joints as I was," the woman reported.

"One day, as Mrs. Kane felt somewhat improved, she unexpectedly asked for paper and a pencil. She had a small table standing by the side of the bed. Placing the paper I handed her on the table she began to write feverishly and kept this up till she had filled some twenty pages with rapid scrawling. I did not know what she was doing until she had finished and handed me the pages. I found that she had written down a detailed story of my life." The woman's mother had died earlier that year, apparently without writing a will. The message scribbled by Maggie Fox Kane not only claimed that a will existed but gave directions to it, in a desk at the home of some friends.

"I wrote at once to my brother," the neighbor said. "He sent a friend to investigate. The family in question said they knew nothing about the missing will but invited him to search the desk and the will was recovered." The woman was not a spiritualist. She said she could not explain what had happened, but it did make her wonder what lay behind the mythology of the Fox sisters—and the destructive pattern of their lives.

Theodore Flournoy had certainly wondered about the stresses of being a working medium, about what in their lives might be real and what might be fantasy and wishful thinking. The University of Geneva psychologist directed his fascination with such questions into such a lengthy study that it eventually grew into a full-sized book, published in 1900.

Flournoy had joined Richet in his investigations of Eusapia Palladino, but the Swiss psychologist was far more interested in another practitioner of supernatural arts, one that he found significantly more credible. For his own case study, Flournoy chose the French medium Catherine Muller, who worked under the pseudonym Helene Smith. The resulting book, *From India to the Planet Mars,* explored Mme Smith's multiple trance personalities and examined her strengths and her weaknesses.

Helene Smith was not a professional medium. She conducted sittings for friends and acquaintances. Like Rosina Thompson and Leonora Piper, she did not charge for her time. Outside of the séances, the medium was a respectable thirty-year-old woman, "beautiful, vigorous, with an open and intelligent countenance," who was liked and respected by her neighbors and who worked for a business firm where, Flournoy said, her ability and integrity had led to her being promoted to a managerial position.

Flournoy had first visited Mme Smith anonymously and had been shocked when she began discussing his family, including some events so obscure that he'd had to write to relatives, checking the accuracy of her accounts. He'd been further shocked when the details were confirmed. He tried to find out where she had acquired the information, where she *could* have acquired it. He found no evidence that she spied on visitors, hired detectives, or used any other obvious methods of cheating. He was left with the notion that she had an unusual talent for telepathy, perhaps comparable to that of Leonora Piper.

What complicated a consideration of Mme Smith's abilities was the dubious nature of the trance personalities through which she communicated her extraordinary knowledge. These "spirit guides," like Mrs. Piper's Phinuit, seemed extremely unlikely to be the afterlife manifestations of actual people and more likely to have sprung from the depths of the medium's own mind. A peculiar assembly of characters jostled for supremacy once Helene Smith slid into a trance. They included a kindly Victor Hugo; a hostile military leader, who would become so angry he would pull the medium's chair out from under her; the doomed French queen Marie Antoinette; a domestically inclined Martian; and the long-ago wife of a Hindu prince.

Flournoy thought Mme Smith's trance personalities were both part of and independent from her possible telepathic gifts. That is, her mind might create them as it struggled to cope with processing the thoughts and needs of other people. But the personalities were undoubtedly created from her "subconscious, memories, scruples, emotional tendencies." He suspected that the characters arose from forgotten experiences in her childhood, resurfacing as the fatherly Victor Hugo or the childish, whispery Marie Antoinette.

Thus, her most exotic séances might result from a kind of mental embroidery, building a small gift into something more exciting. The Martian and the ancient Indians who came calling in her trances didn't impress Flournoy as much as the way she could occasionally peer inside a visitor's head. But most of her visitors felt differently; they were thrilled by this eerie contact with savages and aliens. The trance personalities revealed the mind of the medium; they were evidence of a lonely woman seeking attention and respect for a gift that could—especially if unappreciated—become a burden, possibly an unbearable weight.

"UPON MY WORD, dear Flournoy, you have done a bigger thing here than you know; and I think that your volume has probably made the decisive step in converting psychical research into a respectable science," James wrote shortly before leaving for the south of France. He hoped that more such case studies could be done and that Hodgson's work with Mrs. Piper could be expanded to include some of the analysis that made Flournoy's account so insightful. "Your book has only one defect, and that is that you don't dedicate it to me," James joked, adding somewhat gloomily that in his current state of health he would "very likely die with my great Philosophy of Religion buried inside me and never seeing the light, it would have been pleasant to have my name preserved for ever in the early pages of your immortal work."

But at Richet's chateau, washed in light, soothed by the salt-tinged breeze, James felt his spirits lifting, and with them his health. He began spending less time in bed, more time on the chateau's terraces, wrapped in blankets, tucked into an oversized rocker, letting the day glimmer around him. At Richet's chateau, James ate fish, artichokes, and the stewed lettuce that was considered a health enhancer. On the veranda, he soaked up the sunshine and admired the surrounding fields of hyacinths and violets (grown for export), and his health improved, "*tho* with extreme slowness."

"I have got to this splendid sunshine and out of door life and everything has taken an upward turn," he wrote to Hodgson. James looked forward to sharing Richet's "noble country house" with Myers and his newly discovered medium. And he was eager, already, to feel well enough that he could go home and pick up his discussions with Hodgson about Leonora Piper. "I believe the good days are to come again."

Myers proved enjoyable company, although James found the pretty Evie "rather a spoilt child." He liked Mrs. Thompson, though; she had a quiet dignity that reminded him a little of his favorite Boston medium. In the evenings, he and Myers would disappear with Mrs. Thompson into a small study and test her trance effects.

"The most unfortunate circumstance is that with Mrs. Thompson as with Mrs. Piper, the most striking evidence of her powers is too private for publication," James wrote to a friend. The rest of it was the usual tumbled mix of aphorisms and trivia. She had provided a flood of information about a recently deceased friend, but his "spirit" seemed to dwell inordinately on his walking stick and fur collar and his failing mind before death. James again had to wonder why a returning spirit would be so obsessed with such minutiae as collars and canes.

"We are having the D——l's own time with Mrs. Thompson, Myers' medium here, who is the greatest puzzle out," James wrote to his son, Harry. She induced in him the familiar sense of bafflement, the usual mixture of hope and of doubt. His wife thought he wasn't well enough for the experiments anyway, that spending hours with a medium—and the usual assortment of complications that entailed—was far too stressful for him.

It was almost a relief when the group broke up in March. The Myerses left for Paris. Mrs. Thompson happily returned to her London home. The Jameses, too, were preparing to move on, first to visit Theodore Flournoy in Geneva and then to the German spa for William to be "examined and sentenced" by the doctors. "Your mother is extremely rosy and well," he wrote to his daughter. "She has no complications now that the Myerses and the medium are gone."

FOR JAMES HYSLOP, the hard work was just beginning. He'd finished his masked séances in Boston with Leonora Piper and had marveled at some of the results. Back in New York, he set out to double-check, pick apart, verify, or discard every statement made by the "spirit" of his father during those sittings.

Following up on a séance in February 1900, he wrote, "My Dear Mother: Please to answer the following questions and return this with reply: 1) Have you had a rheumatic trouble either since I saw you, or sometime before? If so, how are you now? 2) Did we have a horse by the name of Jim within your recollection? 3) What became of the horse named Bob?"

And on another tack: "Did father ever speak to you about my theories being strange? If so, do you remember what they were in particular and how he spoke of them?" Hyslop carefully didn't mention that the question referred to a comment about Swedenborg.

A PROPHECY OF DEATH   245

He closed with the formality typical of his family. "Yours as ever, J. H. Hyslop."

His mother answered every query: She suffered from neuralgia but not rheumatism. They had owned horses of those names; Bob had been put down after Hyslop's father's death. And Hyslop's father had thought that his son's ideas were very peculiar indeed, especially on the evening that he explained Swedenborg to his bewildered fundamentalist Christian parents.

"I think he remarked afterwards when we were talking about the conversation that you had some strange ideas," his mother wrote, signing herself, "Yours affectionately, M. E. Hyslop."

The son neatly recorded her answers. He checked them against his own knowledge. "When you remember me to father," he wrote to Hodgson in Boston, "please say to him that he was right and I was wrong about that incident in regard to Swedenborg."

Hyslop found it difficult to express in writing that sense of personal recognition. The turn of phrase, the expressions chosen, had been so like the way his father talked. One evening, Hyslop's "father" had said, "Do you remember what my feeling was about this life? Well I was not so far wrong after all. I felt sure that there would be some knowledge of this life, but you were doubtful, remember you had your own ideas, which were only yours, James." To an outsider, it might sound a vaguely encouraging statement. But Hyslop couldn't count the number of times that his father had told him that, "You have your own ideas. . . . He meant that I was the only one of his children who was skeptical, and this was true."

Over four sittings, Hyslop said the "ghost" of his father had described 205 incidents, of which 152 were true, 37 unverifiable, and 16 false. And while the tally was reasonably impressive, he doubted that it conveyed his own bone-deep assurance that "I talked with my discarnate father with as much ease as if I were talking with him, living, through the telephone."

Mrs. Piper's séances could be a heady experience, leaving one overconfident of opening the doors to immortality. Hodgson warned Hyslop to think of this as a temporary euphoria. Before he became too enthusiastic about accessing the occult, the philosophy professor needed to spend some time on the professional medium circuit.

Hodgson sent Hyslop to check out the cozy little Occult Bookstore, on New York's West Forty-second Street, and observe the working mediums there. A few evenings later, Hyslop was seething with outrage. "And the effrontery of the whole business is one of the most amazing things I ever met," he wrote to Hodgson, although he found himself as annoyed by the "fools who fell for the scam—and paid to hear such nonsense!" as he was with the con artists themselves.

The scam was a variation on the old sealed-envelope ploy. Sitters were asked to write their questions on small blank pieces of paper and wad them into tight, tiny pellet shapes. These crumpled balls were heaped on a table, in plain sight, usually placed in a brass bowl or tray. Mediums appeared to barely approach the pellets, perhaps brushing them with a fingertip, no more.

As Hodgson had warned him, and as Hyslop rapidly confirmed, if one simply kept one's eyes on the mediums' hands, the whole show revealed itself—pellets were palmed, substituted, scanned. Distraction was the key. One famous pellet reader simply lit and relit his cigar, using the motion to palm the pellets, holding the match each time so that it illuminated the message, dropping the pellets back in the dish.

During one sitting, Hyslop became so exasperated that he put a pellet on the table, then declared that he was embarrassed and ripped the paper to shreds, then announced that he'd changed his mind and asked for a spirit opinion on the message anyway. Naturally, as he told Hodgson sarcastically, the sitting was a failure.

"It is rather amusing about the pellet that was torn up," Hodgson wrote him. But after all, genuine psychics didn't need to rely on such showy demonstrations. "It is almost a sure thing, when the writing of names and questions on pellets come into a sitting, that there is fraud."

Hyslop found himself in complete sympathy with Hodgson. He agreed that the commercial medium trade was a scummy business. And he agreed that Leonora Piper was a different entity altogether.

Determined, as always, to be honest in his opinions, Hyslop wrote an essay for *Harper's,* firmly making his case for communication with the dead. Coming as it did from a Columbia University philosopher, the piece attracted considerable attention. "Horribly written," William James commented, "but makes a stronger case for spirit return than anything I've seen."

After researching and analyzing his Mrs. Piper sittings, Hyslop had considered and discounted the possibility of fraud or trickery. He'd been unable to find even a hint of evidence that Mrs. Piper had gone visiting Xenia, Ohio, sent detectives there, or made any inquiries whatsoever. In a community so small and close, he would have heard about that immediately.

That left Hyslop with two possible explanations for such uncanny knowledge making its way into Mrs. Piper's trance state: "omniscient telepathy and discarnate spirits." After reviewing the evidence, he could see no reasonable answer but spirit communication. To start with, many of the facts Mrs. Piper provided about his family were unknown to him at the time. She could not have read his mind for things he'd didn't know. As he'd ruled out a secret intelligence system, the best remaining option, he then concluded, was that his father's spirit had been in attendance.

Hyslop admitted that such a conclusion sounded improbable. Yet science was constantly exploring other improbable places, he pointed out, one might even say "wasting enormous resources upon expeditions in search of the North Pole, or in deep sea dredging for a species of useless fish to gratify the propensities of evolutionists. . . . Why is it so noble and respectable to find whence man came, and so suspicious and dishonorable to ask and ascertain whither he goes?"

New York journalists began packing into Hyslop's talks, hoping for further signs of lunacy. "It was funny to study the newspaper reporters," Hyslop wrote to Hodgson after one talk. "They came there as usual to watch and hear cranks" and left disappointed by his pedantic description of his work.

His detachment didn't last long, though. There were, after all, more than two dozen fiercely competitive city newspapers. Dull copy didn't sell. Hyslop's cautious way of describing his research wasn't nearly as interesting as the journalists wanted it to be. More than one journalist jazzed up a story by writing that Professor Hyslop proposed to "scientifically demonstrate the immortality of the soul," perhaps in the next few weeks.

It dismayed Hyslop to learn that his scholarly peers were willing to take such ridiculous newspaper accounts seriously. Many didn't care that Hyslop had been misquoted on his intentions. They cared that he had given unwarranted support to psychical research. At Columbia, James McKeen Cattell, still simmering over his debate with William James on the same subject, led these hard-liners. Outraged to find his own university associated with spiritualist nonsense, disappointed that any faculty member would call spirit communication as worthy a pursuit as Arctic exploration,

Cattell openly demanded that the university president order Hyslop to abandon his crusade. Rumors had it that Cattell also met privately with Columbia's president, demanding that Hyslop be censored and then fired.

Hyslop found himself in a fight for survival. He had some support from the younger faculty, but several told him they were afraid of Cattell's influence on their own careers. He decided to court more established professors and administrators to gather more support against "Cattell's insolent interference with me. . . . I shall make a very fight if it comes to the point where it is necessary and Columbia will not forget it for 50 years," he wrote to Hodgson.

Hodgson encouraged him to stay calm. "It would be pretty absurd for the authorities of the college to make any move on the ground of newspaper reports about what you said at meetings. . . . be sure and take everything quite coolly." But Hyslop had never been cool-natured, and he couldn't help but worry. He was a married man. He had two children. He hadn't expected that a scientist at his own university would regard a difference of opinion as something to be suppressed and punished.

Columbia's administration, anxious to avoid a public fight, suggested that the whole affair would blow over if Professor Hyslop would practice more discretion. It seemed his only option. Hyslop told Hodgson that he'd canceled scheduled talks on psychical research and planned to immerse himself in classroom duties.

Hodgson expressed sympathy, but he was angry as well, dismayed by the vindictiveness of scientists in general and Cattell in particular: "He hasn't a true scientific spirit at all, nothing but veneer, and it has always seemed to me extraordinary that he should have got to the positions he occupied."

If Cattell had done an objective investigation of events, or even shown the slightest interest in the truth, he would have realized that the stories were simply journalistic exaggerations, Hodgson continued. Someday, Hodgson hoped, Hyslop would be able to expose Cattell's own "unscientific attitude" regarding psychical research and turn the tables around.

HENRY SIDGWICK HAD not been not quite well for the last few months. Tired and achy, he had at the urging of his doctor gone to London in early May to consult a surgeon.

The diagnosis shocked him beyond measure; Sidgwick had all the signs of a fast-spreading cancer. The surgeon wanted to operate quickly but warned that surgery would only delay death briefly. For two weeks, Sidgwick quietly stayed home with Nora, seeking and giving comfort. By late May, though, the surgery was scheduled, and he began telling friends and family members.

"A terrible day," Sidgwick's brother wrote in his diary after their conversation.

Sidgwick wrote to Myers in France, apologizing for sending the bad news by mail, letting him know that he hoped to survive the surgery but not the year: "Life is very strange now: very terrible: but I try to meet it like a man, my beloved wife aiding me. I hold on—to try to hold on—to duty and love; and through love to touch the larger hope. I wish now I had told you before, as this may be farewell. Your friendship has had a great place in my life, and as I walk through the Valley of Shadow of Death, I feel your affection.

"Pray for me."

Sidgwick's doctors operated in mid-July, on his sixty-second birthday, removing as much of the cancer as they could. As soon as he was able, Nora took him to convalesce at the country home of the Rayleighs.

In late August, Nora summoned other family there. It was time to say good-bye: "We now have no hopes for Henry, but that the growing weakness, which he bears with unbroken patience and simplest unselfish fortitude, may soon reach the natural end he so desires," his brother Arthur wrote to a friend.

Sidgwick died on August 28, 1900. He was buried in the village churchyard near the Rayleighs' home. To the end he cherished his doubts about God, his sorrow that he'd failed to prove the existence of a higher power. He begged not to have a church service over his grave. He had written down, instead, a few brief lines for a minister to read: "Let us commend to the love of God with silent prayer the soul of a sinful man who partly tried to do his duty. It is by his wish that I say over his grave these words and no more."

It would be an honest good-bye, Sidgwick told Nora; it would be a fittingly moral end to his life. She knew he was right. But she couldn't make herself let him go so simply. In the end, he was buried with all the pomp and ceremony and calls to faith of a traditional Church of England service.

"Everything seems left undone in this world," James wrote to Nora Sidgwick in early September. He'd come to believe, he told her, that her husband had kept psychical research sane and steady. James did not know who could replace Sidgwick in that role.

James and Myers were both sick again, each losing ground after leaving the gentle life at Richet's chateau. James was en route to Germany for yet another round of medical treatment. He knew that Nora had decided to leave England for a while. She planned a trip to Egypt, where some of her students were working on an archaeological dig and where she might revisit comforting memories of the pleasures of doing math along the Nile.

"Dear Mrs. Sidgwick, you have no idea how many of us mourn with you in this bereavement or what an impression of flawlessness in quality your husband left by his person on all those who knew him, and by his writings on those who never saw him," James wrote. "A spotless man, a wise man, a heroic man."

WILLIAM JAMES NOW feared he might never again be well. Month after month, country after country, doctor after doctor, he could not seem to shake off his pain and lethargy. In despair, he wrote to President Eliot at Harvard and offered to resign his faculty position. Instead of accepting, the university extended his leave, Eliot assuring James that he was a philosopher-psychologist worth the investment.

Relieved and grateful, James decided to winter in Italy, hoping that its famously balmy climate would restore the good health achieved on the French Riviera. In Rome, James began a more aggressive treatment as well, injections of compounds taken from the lymph glands, brains, and testicles of goats. Its advocates guaranteed that the murky serum delivered animal health and vigor.

As December arrived and the year drew toward a weary end, James wrote to Myers that his health was on the mend; "my brain power is almost nil. But the arterial degeneration, *mirabile dictu,* does actually seem to be taking a back track." He urged Myers to come to Rome and try the therapy. Even more urgently, James implored Myers to forget the death prophecy from Mrs. Thompson's séances—which had come up again at the chateau—

to let go his promised reunion with Annie Marshall, and to put his energy into all the good life and work yet to come. "I do hope & trust, dear Myers, that your health is keeping up, and that in spite of devils, prophets, mediums and imps, you are to live long for the comfort of your family, the delectation of your friends, and the instruction of the world," he urged.

Myers wrote back, agreeing to come to Rome and try the recommended injections to please his friend, although he doubted that they would save him. Myers thought rather that he would die anyway, and then "return as a cross between an old goat and a guardian angel."

Privately, James feared that Myers wished to die, was not really fighting his illness, that "his subliminal is, to put it brutally, trying to kill him as well as it can," as he wrote to friends at the SPR offices in London. In genuine concern, James suggested the SPR hold a séance with the express goal of getting messages to "neutralize the prediction."

Before January 1901 was half over, though, James thought he knew why Myers found the image of Annie so appealing. The dead lover probably glowed as pure gold in contrast with Myers's living wife.

Terminal illness in a loved one did not bring out Evie's best qualities. Nervous and fretful, she longed openly for her house, her own servants, her children and friends, gossip and conversation. Evie wanted company and comfort; she followed James around like a desperate puppy. "That intolerable babbler, Mrs. Fred Myers has interrupted me by coming and talking and my mind can't discharge the echoes of her voice," he grumbled in a letter to his son Harry.

James had no opportunity to gather life-affirming messages from the spirit world for his ailing friend. Myers had contracted double pneumonia by the time he arrived in Rome; he was struggling to breathe, fighting such intense chest pain that James was called upon to act as a doctor as well as friend. He administered morphine instead of messages of renewal.

With a more practiced Italian physician on the case and visitors banned from the sickroom, the Jameses were left to entertain Evie and to send Myers notes of encouragement: "I think of you all the time patiently undergoing this ordeal, and my sense of human nature's elevation rises," James wrote. He told Myers not to worry at all about his unfinished projects. Even as he stood by, James was editing a paper Myers had prepared on

Rosina Thompson, which described a visit from Annie Marshall (cloaked in anonymity) and a detailed conversation with a spirit who sounded a lot like Edmund Gurney.

James also promised to see Myers's book-in-progress, *Human Personality and Its Survival of Bodily Death,* published—even if he had to finish writing it himself. But he assured Myers that he expected his friend to be shortly back at work, the crisis "just a memory." Myers didn't argue with such optimistic predictions. He lay without complaint, asking only that his favorite poetry and philosophy books be read to him. Finally, with death imminent, the doctor allowed him company. His wife and friends sat by him, reading loudly over the harsh gasps of his breath.

He died on January 17, 1901, not six months after they had buried Henry Sidgwick. "His serenity, in fact his eagerness to go, and his extraordinary intellectual vitality up to the very time that the death agony began, and even in the midst of it, were a superb spectacle," James wrote to Nora Sidgwick.

The only odd thing was the death itself. Myers seemed to have choked to death, which was highly unusual for the type of pneumonia that he had. Both James and Myers's doctor in Rome were puzzled by it, the Italian physician saying the fatal illness "behaved in a way that he had not seen in 1000 cases."

It led James to ponder, once again, on the power of that prophecy, that message from Mrs. Thompson that Annie Marshall was waiting for her Fred, expecting him to join her soon. Had it been a rare instance of clairvoyance, or, as he suspected, had it come true because Myers had willed it to do so?

"IS THERE GOING to be any difficulty about poor Mrs. M? I mean about the degree of knowledge or ignorance she may possess of the A-Control," wrote William James to Oliver Lodge, in a carefully cryptic exchange in March 1901.

Their private discussion occurred at a time when attention was focused on a much more visible, news-making death. On January 22, five days after Myers succumbed, England's Queen Victoria had died at the age of eighty-one, ending a reign that had lasted more than sixty years. Her eldest son had since been named Edward VII, king of the United Kingdom of Britain and Ireland, and Emperor of India.

Lodge and James were more concerned with the legacy of the late Fred Myers. The "A-Control" was the spirit of Annie Marshall, carefully documented by Myers, in sittings with both Mrs. Thompson and Mrs. Piper. In addition to those transcripts, their friend had also written a brief, privately published autobiographical monograph, which told of his longing for Annie and his efforts to find her.

"It is a delicate business, that; also for the children," James continued. "Inevitably every thing will leak out, unless there should be a conspiracy of silence more efficiently carried out than seems possible." He recommended being honest with Evie from the outset. Lodge was unconvinced. He hoped to avoid the confrontation, believing that they could continue to hide Myers's other life from his widow.

A larger challenge was to ensure that psychical research continued, despite the deaths of so many of its champions. James warned that Evie— "a foolish kind of woman"—was unlikely to make things easier and would probably consume far too much of their time and energy. Already, in fact, she was working herself into a state. James felt bombarded by her letters telling him that she didn't like Lodge, didn't like Hodgson (who had been assigned to finish Myers's book), and didn't like the way the SPR was preserving her husband's memory.

Evie herself was trying to answer an overwhelming stack of letters to "my Fred." The writing seemed endless, she told James. Her life was terrible, and she suffered from the "physical anguish" of loss. And then—as if to prove that they should have been candid after all—in going through her late husband's papers, Evie found an early draft of that hidden autobiography, "Fragments of an Inner Life."

She wrote immediately, and slightly hysterically, to both Lodge and James, berating them for concealing it, and demanding that any copies be destroyed, along with all records of the Annie Marshall sittings, whether they seemed solid or not. James and Lodge might consider it evidence of spirit contact—but what did that matter if it was also evidence of a failed marriage?

. . .

AS LATE SUMMER came in, with its lazy golden afternoons and slowly fading gardens, as the Jameses were returning to Cambridge from Europe, the battered psychical research group received more bad news. Leonora Piper had announced that she was calling it quits as a medium.

She'd spent enough time in enforced social isolation, feeling a kind of freak, giving up her days to the blurred reality of a trance. She wanted picnics by the Charles River, teas with friends, Sunday mornings at church, the long-deferred prospect of a normal life.

Mrs. Piper chose not to confront Hodgson directly; he was far too persuasive a debater. So she issued her declaration of independence another way. She volunteered an interview to the *New York Herald,* declaring that she "would never hold another sitting with Mr. Hodgson, and that [she] would die first." All her bottled-up fears and uncertainties came spilling out. She didn't know what happened—or what happened to her—in the trances: "I am inclined to accept the telepathic explanation of all the so-called psychic phenomena," she said, "but beyond this I remain a student with the rest of the world."

She felt no more than a scientific test object, an "automaton going into what is called a trance condition" for purposes of investigation. Eighteen years of study and unsolved mystery felt like enough of her life. She was giving it all up, and she planned to devote herself to "more congenial pursuits."

Shortly before setting sail for Boston, James had assured Evie Myers that her late husband's unfinished masterpiece, *Human Personality,* was coming "rapidly forward." On arrival, he found that promise to be an inadvertent falsehood. The book rested in Hodgson's care, and that worried psychical researcher had barely looked at it as he attempted, fruitlessly, to reason with Mrs. Piper.

James plunged into the business of mending fences. He began by telling Hodgson that he needed to work on his manners. As Oliver Lodge complained, their intrepid Australian investigator rarely bothered with the niceties of social behavior, "being absolutely fearless and uncompromising in expressing what he believes to be the truth; or, for the matter of that, the lie." In the case of Mrs. Piper, Hodgson admitted that he'd become increasingly remote in order to fend off any suspicion of a close relationship. In their recent trance sessions, he'd barely said hello before beginning a sitting, and left with a cool farewell at the end.

Not always the most diplomatic of men himself, James "remonstrated" with Hodgson for treating Mrs. Piper as a somewhat balky machine and secured from him a promise to be more considerate. James then made a formal apology to Mrs. Piper, assuring her that all the SPR members held her in high regard, urging her to take into account the importance of the work, and begging her to reconsider.

Leonora Piper already regretted her flash of rebellion. The New York papers had made her look like a fool, she thought—a weepy, hand-wringing sort, which she wasn't. Further, the headlines had declared that she'd recanted her entire career, that she had practically announced herself as a fraud. She hated reading accounts so entirely untrue.

After thinking it over for several days, Mrs. Piper gave another interview—this time to the *Boston Advertiser*—criticizing the *Herald* for labeling her interview a "confession." She was returning to her work with the ASPR and Dick Hodgson. The only thing she wanted to confess to was puzzlement and frustration: "My opinion is today as it was 18 years ago. Spirits may have controlled me and they may not. I confess that I do not know."

VACATIONING IN THE Adirondacks, thinking his position at Columbia was well secured, James Hyslop had the misfortune to encounter a fellow guest who reminded him all too well of his nemesis, James McKeen Cattell. The astronomer Simon Newcomb, the difficult first president of the ASPR, was relaxing at the same resort.

Newcomb made the mistake of rather mockingly inquiring about the latest in spiritual study. Hyslop, seething with suppressed outrage, allowed himself to vent for nearly three hours. As he wrote to Hodgson, "I poured experimental telepathy into him and then the Piper incidents until he was ready to cry enough and at last told him that he could choose between accepting telepathy or something worse." The argument ended when Newcomb—to Hyslop's frustration—merely walked away.

The ill will stirred in that encounter seemed an omen of events to follow. Hyslop's wife, Mary, returned from the vacation with meningitis. She was dead in three days, "a terrible shock," Hyslop said, as he grappled with being the single father of three children.

Now, at the worst time, Hyslop began to again feel backlash from the antipsychical research camp at Columbia. Cattell still wanted his dismissal. Out of patience, the university president attempted a compromise. He shifted Hyslop out of his accustomed classes and into an intensive schedule of teaching advanced metaphysics. The shift, with the extra preparation time involved, greatly increased Hyslop's workload. By fall's end, he was staggering with weariness. He'd developed a racking cough, which "I soon discovered . . . was tuberculosis and that it had been precipitated by nervous prostration."

Hyslop requested and was granted a leave from Columbia. He sent his children to stay with relatives and checked himself into a sanatorium for consumptive patients in Saranac Lake, New York. To his surprise and relief, he found the cure effective—not just in restoring his health, but also in improving his outlook. He took long walks, did deep breathing exercises, and gave up both coffee and alcohol. He even tried a recipe—sent to him by the sympathetic Hodgson—for making pine-bark tea from tree scrapings.

By the fall of 1902, he thought himself well enough to take up his duties at Columbia. Once again, though, he was given a grueling schedule, and once again Hyslop tumbled into illness. He dropped eighteen pounds in six weeks—pounds that he could ill afford—and coughed his way through classes. When the semester ended, Hyslop resigned from Columbia. He believed that he needed a physically active, outdoorsy life to keep the tuberculosis in check. He decided to join some friends who were starting a gold-mining company in the Green Mountains of Vermont. He sent his regrets to only one person, writing to Richard Hodgson and expressing his sorrow at abandoning psychical research and a friend.

WHY ARE SOME people's minds so open to faith and belief, and others locked tight against those ideas? Why does a god appear necessary to so many cultures? To William James these were fundamental questions, and he asked them directly in his 1902 book *Varieties of Religious Experience: A Study of Human Nature.*

To the question of individual differences in faith, James applied Myers's concept of a subliminal self. Some people might be almost entirely focused on the conscious world, unable to detect any sensation of an otherworldly reality. Such individuals might well become scientists or pursue other fields based in logical deduction. Perhaps other people, more naturally open to subconscious experiences, would be more inclined to accept miracles or spiritual powers. Perhaps the varieties of religious experience were based in a kind of scientific reality, in the varied ways that people's minds operated, the alternate realities that they perceived in forming their worldviews.

By this analysis, a Leonora Piper might be unusually receptive to psychical phenomena and ready to accept the notion of powers beyond human control. A James McKeen Cattell might be unusually closed to any such experiences, thereby finding the supernatural or any spiritual notions to be unfounded and illogical. James defined himself as a "piecemeal supernaturalist," demanding better evidence of the spiritual realms but finding "no intellectual difficulty in mixing the ideal and real worlds together."

From that perspective, James went on to raise—and to challenge—a theory popular among Darwinian scientists, which argued that "religion is probably only an anachronism, left over from an earlier stage of human evolution." Known as "the survival theory," it stated that primitive man, in order to cope with a hostile environment, needed explanations that gave reason to life with its all its grief and struggle, denied that tragedies were merely random events, soothed with a promise of personal interest by the powers above.

But as the survival theory had it, civilized man now knew better, and godly explanations were no longer required. The new science found ridiculous the notion that a God capable of creating a universe would cater to the needs of each short-lived individual occupying one meager planet. As James described this modern version of a deity, "the God whom science recognizes must be a God of universal laws exclusively, a God who does a wholesale, not a retail business. He cannot accommodate his processes to the convenience of individuals."

Thus, both traditional religions and unorthodox ones such as spiritualism could be seen as vestiges of an earlier stage in human evolution. The theory predicted that the human race would eventually cast off that primitive need entirely. It reflected the hope of many modern scientists that John Tyndall's assertion—that science would supersede religion as the way to understand life and its limits—was on its way to being a twentieth-century reality.

William James had no such hopes, nor any fondness for this rational future that so many of his academic peers eagerly anticipated. The survival theory, he wrote, ignored the fact that civilizations come and gone had also been arrogantly sure that they possessed the one Truth above truths. He thought it a mistake to dismiss the ideas of history simply because they didn't fit current scientific methodology.

As Myers's concept of subliminal consciousness emphasized, people didn't fully understand yet what was inside their own brains, much less the world without. Even as an accredited academic, James couldn't make himself believe that "the boundless universe" was so simple as to be easily measured by mortal men. Even the supreme scientific confidence of the new century could not alter that position: "Humbug is humbug, even though it bear the scientific name, and the total expression of human experience, as I view it objectively, invincibly urges me beyond the narrow 'scientific' bounds."

EVIE MYERS WANTED every trace destroyed, every scrap of evidence, that her husband had been infatuated with a spirit. In that hated autobiography, "Fragments of an Inner Life," Myers had actually counted the days with and without his beloved Annie. "Only on 426 days of my life—now numbering more than 18,000 days—did I look upon her face; but that was enough." Even worse, following his epiphany during the Mrs. Thompson sittings, Myers added euphorically that "love has surmounted the sundering crisis" and that he could hardly wait to view that beloved face once more.

Myers had given privately printed copies of "Fragments" to a few close friends. Evie wanted James, Lodge, Hodgson, and anyone else in possession of the humiliating document to turn his copy over to her. She asked William James to oversee the recall, beginning with Hodgson, who was such a difficult person; and she also asked that he please visit Lodge when he was next in England.

She couldn't talk to Lodge herself. She didn't trust him; she didn't trust his wife; she didn't even trust his children. She hoped James would persuade Lodge to give up his copy in the name of his friendship with her dead husband, who would have been "the *last* person to wish to harm his wife and children—and this would be the inevitable result," if the autobiography gained wider circulation.

Lodge was in an assertive mood. In early 1902, King Edward had knighted him for his scientific achievements. While waiting at the palace to officially become Sir Oliver, he'd struck up a friendship with a writer also awaiting the knighthood ceremony, Arthur Conan Doyle, author of the Sherlock Holmes mysteries. Lodge and Conan Doyle had continued their acquaintance, and their discussions of spirit communication. The talented mystery writer would eventually become a notable spiritualist and author of a two-volume history of the subject, which he dedicated to Lodge as "a great leader in both physical and psychic science."

Sir Oliver Lodge now headed the physics department at the newly established University of Birmingham. He'd made five demands before taking the Birmingham job: a research laboratory, the retention of his secretary and two assistants, an endowment fund, protection from "mundane" aspects of research projects, and no interference with his interest in psychical research, "although I knew it would be unpopular."

In his spare time, Lodge was engaged in a patent battle with the Marconi Radio Company, demanding (ultimately successfully) that the company honor his early work on radio receivers. He'd taken to tinkering with automobile mechanisms and invented a new kind of spark plug for internal combustion engines—the power behind the horseless carriage. Two of his sons were drawing up plans for a Lodge Motor Plug Company.

As Myers's friend, Lodge wasn't willing to abandon his friend's autobiography. He thought the book a lovely, eloquent thing, laced with some of Myers's best poetry. True, many of the poems were love songs to Annie Marshall, but in Lodge's opinion, Evie's actions would only serve to erase Myers from his place in history. The best Lodge was willing to offer was a compromise. The society would, for now, publish only a few of the better poems, as a booklet called "Fragments of Poetry." But, Lodge insisted, Meyers's friends would also save the manuscript as a record of a friend, of a love affair, of faith in the possibility of life after death.

In February 1903 the publisher Isaac Funk—famed for the *Funk & Wagnall's Dictionary*, less well known for his ongoing curiosity about spiritualism—decided to visit a private medium in Brooklyn.

At first the woman, a sixty-eight-year-old widow, seemed unimpressive, maybe a little unstable. In trance, she seemed to represent a veritable crowd of spirits—male, female, drawling southerner, twangy westerner—and Funk, bored, decided that he was observing some kind of strange split personality disorder. As he wrote later, the idea of mental dysfunction was fixed in his mind, "up to the time that I had the singular experience which I give below."

The medium suddenly announced that a spirit had come who was concerned about an ancient coin, called the Widow's Mite for its tiny size and minimal value during its day.

"This coin is out of its place and should be returned," the Brooklyn medium insisted, adding sharply that the visiting spirit looked to Dr. Funk to return it.

"What do you mean by saying that he looks to me to return it?" Funk demanded irritably. He possessed no coin belonging to any spirit.

The woman repeated the demand.

Funk fumed for a while and then, gradually, recalled that when his company was making the first edition of *The Standard Dictionary*, nine years earlier, he had borrowed an old coin, which had been called the Widow's Mite, to copy for an illustration.

But he'd given it back.

"This I promptly returned," he repeated.

"This one has not been returned," came the reply. Funk was advised to look for it in a large iron safe, in a drawer, under a pile of papers.

When he returned to his offices in Manhattan, Funk queried his longtime business manager, who recalled the coin with some trouble, and also that it had been returned years before. The company cashier told him the same.

The cashier agreed, however, to search the iron safe in the business office, with his assistants for witnesses. In a small drawer, in a dirty envelope, pushed under a muddle of papers, they found the coin.

Funk mailed it back to the original owner and received a letter from the man's son, saying that his father had died years back, but he too had thought the coin returned. "As executor of my father's estate, I felt so certain that this coin had been returned that it never occurred to me to make inquiry of you whether it was in your possession."

Funk was an organized man, a list maker by nature. He drew up four possible explanations: fraud, coincidence, telepathy, or spirit communication. To find the correct answer, he decided on two actions: to tell the story publicly, seeking comment, and to consult with experts.

Funk wrote to forty-two scientists, editors, philosophers, and other scholars, describing his "Widow's Mite" incident and asking them which of his four possibilities made sense. He queried James Hyslop, William James, Alfred Russel Wallace, Sir William Crookes, and professors of physics and philosophy at Yale, Princeton, the University of Toronto, the University of Wisconsin, Vanderbilt University, Columbia University, the University of Michigan, Cornell University, and a smattering of other institutions across Canada and Europe.

The resulting answers didn't resolve the question, but they did neatly reflect the schism in thought between psychical and traditional researchers. "Spirits," replied Wallace and Crookes. "Possibly spirits," answered Hyslop and James. The rest of the scientists queried, to a man, voted for either fraud or mental illness.

"A batch of reporters came after me about the Dr. Funk case," complained Hodgson to Hyslop, adding that he wished psychic euthusiasts would keep their alleged test cases out of the newspapers. Thanks to Dr. Funk's well-known name and the peculiar nature of his experience, the

Widow's Mite was front-page news in New York, jostling for space with accounts of the trans-Pacific telephone line connecting Canada and Australia, President Roosevelt's successful move to take over the Panama Canal, and an extraordinary decision in Australia to allow women to vote (making it and New Zealand the only two countries thus far to permit such electoral inclusiveness).

Hyslop's gold-mining venture in Vermont had not been a financial success. But his tuberculosis was once more in abeyance. He was stronger and more determined than ever to prove that he had been right about spirit communication and that his enemies at Columbia had been wrong. He rented a small apartment on New York's Upper West Side, supporting himself with some family money, some writing jobs, lecturing, and the occasional investigation for Hodgson.

Hodgson, meanwhile, tried to finish Myers's book while continuing his work with Mrs. Piper. He did his best to ignore Evie Myers's frequent letters accusing him of going about it in the wrong way. He was delighted to have Hyslop again as a friend and ally. A colleague described them once as a curious pair: "Hodgson with his lithe athletic frame, a perfect dynamo of mental and physical energy; Hyslop, inactively tubercular, frail of physique, but with tremendous high tension enthusiasm, appearing to almost consume physical vitality as a flame consumes the candle."

But they were alike in their single-mindedness; "It was hard to say which of them had a greater hatred of sham, hypocrisy and academic cowardice." Hyslop and Hodgson shared thoughts and ideas and complaints, letters between them going from Boston to New York discussing fraudulent mediums, pig-headed scientists—and the mistakes made by the British and American societies for psychical research.

Hodgson thought the SPR's British leaders were too willing to believe. They'd stumbled badly with Eusapia Palladino; he wasn't sure they wouldn't fall again without his help. Recently he'd written several stiff criticisms of Rosina Thompson, only to have them edited out of the British journal. "In fact, the more I come to think of it, the more I feel like that old woman who has been so often quoted, who said that she 'wasn't sure that anybody at all would be saved but herself and her husband Sandy, and there were times when she was nae so sure of Sandy!'"

The Americans stood at the other extreme. They were "morbidly afraid" of belief, as Hodgson put it. The ASPR had turned away countless interested supporters on the grounds that they weren't skeptical enough. No wonder the organization had stayed small and poor. Hodgson now owed his secretary, Lucy Edmunds, almost $900 in back pay, which had accumulated over four years, and he himself had spent $600 of his own money just to keep the office supplied with up-to-date publications.

He wasn't complaining on his own behalf. True, he was still unmarried, still living in two small rooms on Charles Street, but the rooms were stacked, floor to ceiling, with his greatest indulgence: books—poetry, philosophy, novels, science. For fun, he talked to his pet parrot, which he also occasionally brought with him to the Tavern Club, where he went almost every night for dinner and company. He played handball, fished, hiked, swam in the Atlantic whenever he could. He didn't want for money himself—just to fund the work.

Perhaps, Hyslop suggested, the psychic investigators should stop doing inadequate, underfunded investigations. Perhaps they should save, build up an endowment, wait till they could afford to do it right.

It would be wonderful to have the money, Hodgson agreed, but he didn't think they could afford the time that would take. He didn't want to lose even the small momentum the group had achieved. They had lost Gurney, then Sidgwick, then Myers; they couldn't spare a single warm body. Hodgson would work harder, that was all.

WILLIAM JAMES HAD hoped that Fred Myers's book on subliminal consciousness would be "epoch making." But *Human Personality and Its Survival of Bodily Death,* published in late 1903, almost two years after Myers's death, proved more complicated than that—and less successful.

Myers's ideas had stretched beyond his fundamental concept of the subliminal mind to a far grander proposal, that the connections between conscious and subconscious, subliminal and waking mind, were coordinated by the immortal soul. This meant, as Myers ruefully acknowledged, that to accept his bigger theory, one had to accept the notion of a soul and of its survival after death. In the modern age, simple acceptance would not do, he continued; the question at hand was whether the existence of an immortal soul could be proved. "In this direction," he wrote, "have always lain the gravest fears, the farthest-reaching hopes, which could either oppress or stimulate mortal minds."

In the twentieth century, Myers said, the only way to verify immortality was through the way of modern Science—"dispassionate, patient, systematic," careful, and much to be admired: "Science works slowly on and bides her time, —refusing to fall back upon tradition or to launch into speculation, merely because strait is the gate which leads to valid discovery, indisputable truth."

The only problem was that science, thus far, had chosen not to investigate the question of immortality. Myers believed this omission could be laid to the "resolutely agnostic" view of scientists of the day, which he summed up, "We do not know and will not know." But this recalcitrance, he believed, would be overcome as psychical researchers like himself found inarguable evidence to the contrary. "It is my object in the present work— as it has from the first been the object of the Society for Psychical Research, on whose behalf most of the evidence here set forth has been collected,—to do what can be done to break down that artificial wall," he wrote shortly before his death.

What he could not have foreseen when he composed *Human Personality* was that the evidence that Myers considered strongest—the séances in which Annie Marshall appeared, his many sittings with Rosina Thompson— would not give support to his published argument. His wife had many pertinent records destroyed; more than that, she had refused to allow Hodgson to mention them in his edited version of the book.

In his review of the book for the *Proceedings of the Society of Psychical Research*—one of the more positive published assessments of *Human Personality*—James acknowledged that the "ill-defined relations of the subliminal with its 'cosmic' environment" undermined Myers's case for immortality. At best, he thought, Myers had just managed to demonstrate that possible supernatural events ought "just like other events to be followed up with scientific curiosity." But James emphasized that Myers's willingness to tackle contradictory and confusing subjects should be considered a strength rather than a weakness: "Nature is everywhere gothic, not classic. She forms a real jungle, where all things are provisional, half-fitted to each other and untidy." It was only by acknowledging the messiness of the life itself that a picture of reality could be drawn.

Perhaps others preferred their view of the world to be more orderly, the universe to be delivered in a quantifiable package. James didn't believe in that tidy view of existence for a minute. By accepting the wonderful complexity that Myers sought to portray, he would write, "although we may be mistaken in much detail, in a general way, at least we become plausible."

11

A FORCE NOT
GENERALLY RECOGNIZED

*O*N AN ICY NIGHT late in the fall of 1905, Dick Hodgson hurried with friends across the Boston Commons. The ground crunched underfoot; the night glittered around them, silver-frosted with stars. Suddenly, Hodgson stopped, tilting his head back to study the shimmering sky.

"Sometimes, I can hardly wait to get over there," he said, his right hand tracing a route across the starry pathways. It was so difficult to prove immortality—or at least to convince others that one had—while here on Earth; he now thought, he might accomplish more when he arrived in the spirit world. "I am sure that when I do, I can establish the truth beyond all possibility of a doubt." For a moment, his voice sounded almost wistful: "But I suppose I'm good for twenty years or more at least."

"At least," his friends agreed, laughing, eyeing his bright face and muscular stance.

As always, Hodgson seemed to thrive on his high-energy life. Although he obsessed about the work, he tempered it with leisure pursuits that included his reading, games of handball, socializing, vigorous vacations. Over the previous summer, he'd fished with friends in Maine and hiked in the Adirondacks. In the early fall, he'd spent a happy vacation with the Jameses at their country home in New Hampshire.

"Hodgson left us this morning after a visit of ten days," James wrote to Flournoy. "It is a pleasure to see a man in such an absolute state of moral & physical health. His very face shows the firmness of a soul in equilibrium—another proof of the strength which a belief in the future may give one!"

Mrs. Piper, though, seemed to be faltering. Her husband, William, had died early in the summer. She'd withdrawn into sadness; her sittings had acquired a strangely dreamlike quality. James thought them vague and anemic, as if she'd lost interest, was no longer concentrating.

Even the G.P. personality seemed to be slipping away. In a recent sitting, G.P. had warned Hodgson that their time together in the lamplit quiet of Mrs. Piper's parlor would not continue long. Some days, James told Lodge, he longed for the croaky voice and conniving ways of Dr. Phinuit. Both he and Lodge worried that Hodgson hovered over the Boston medium, overmanaged her so that she functioned in a kind of permanent stress state. Her primary controls these days—new personalities who went by the names of Rector and Imperator—seemed to function entirely to protect her from strenuous demands, making her "inaccessible," as Lodge put it.

James had steered a raft of alternative investigations in Hodgson's direction—a magnetic healer, a teenage girl who practiced "automatic piano playing," an Irish-American dwarf who left streaks of light on any paper he touched—but, to his frustration, none of them had served to lure the investigator away from Mrs. Piper and her mysterious ways.

HAVING RETURNED FROM Egypt determined to pick up her life again, Nora Sidgwick stacked her plate high. She started on a book about her late husband, a combination of memoir, collected letters, and biography. She went back to her duties at Newnham College. She accepted an invitation from her brother-in-law, Lord Rayleigh, to do some tricky mathematical calculations needed for his research.

Rayleigh had retired from his teaching position at Cambridge but was continuing to run experiments in a laboratory he'd built at the family seat, Terling Place in Essex. He was focusing on electric and magnetic problems, the traveling of electric currents and the ways certain materials stubbornly refused to carry those currents. His reputation for nonstop investigation was so established that King Edward VII greeted him at a reception with: "Well, Lord Rayleigh, discovering something I presume?" For his illuminating work on atmospheric chemistry, including the discovery of argon, Rayleigh had received the Nobel Prize in Physics in 1904, the fourth year that the awards were given.

Rayleigh still maintained an interest in occult experiments as well, keeping his membership in the Society for Psychical Research (and eventually becoming president in 1919). For Rayleigh, it was less a matter of conviction than principle. Like other physicists in the SPR—Oliver Lodge and William Barrett—he believed that science was the best tool known. It should always be used for exploring difficult questions, even questions of spirit life and supernatural powers.

Theirs could safely be called a minority view in the research community. Barrett, who followed Lodge in assuming the SPR presidency, noted that prejudice against the society's work seemed to be dying down *except* among scientists, citing two primary reasons for that resistance. First, occult phenomena were not replicable; "the phenomena cannot be repeated at pleasure (any more than a shower of meteorites can)." Second, the strangeness of the subject and the peculiar personalities involved acted as an abrasive on the ordinary scientist's sense of sanity.

THE WORD *peculiar,* indeed, could be applied to almost anyone who dabbled in the occult. One needed only to observe the latest rage in New York mediums, the Reverend May Pepper, who presided over the Church of Fraternity of Soul Communion in Brooklyn.

Pepper charged 25 cents for entrance to the temple. Once the pews filled, she ascended a platform, her long dark hair flowing over her black robes, and asked her assistants to tie a linen bandage over her eyes. She would then "read" letters brought forth by those in the audience.

One memorable evening, described in mocking detail by attending journalists, a young man stumbled up to the lectern bearing not a letter but a whole roll of paper covered with writing.

"Your coming was heralded to me," the medium proclaimed. She'd seen a vision of an Indian reaching out toward him. "Tell me, was there ever an Indian in your family?"

"My great-great-great grandfather fought Indians," the young man answered. "His name was Montcalm."

"Ah," said Mrs. Pepper. "I see an old man in Continental uniform. There is an Indian beside him. The old man is reaching out for you. He says his name is Gen. Montcalm. Tell me, did not Gen. Montcalm leave something which was lost and which you have never found?"

"Yes, yes," the man replied; his mother had been searching for years, following the rumor of a hidden family treasure.

"Never mind about mamma," Mrs. Pepper replied. "The General says he wants a man to do the hunting."

From the journalistic point of view, the story only got better from there. The Anti-Fraud Society of Manhattan had stationed members in the church, who began to make fun of Pepper's visions. Her supporters then came to her defense, detonating an explosion of fistfights that spread across the church.

When the melee ended, spectators discovered that pickpockets had been busy while the battle raged around them, and a multitude of wallets and purses had gone missing. The event was front-page news. Pedantic reports of the psychical research community, meanwhile, rarely merited a mention in the popular press.

William Barrett might complain that scientists couldn't see past the peculiar nature of the subject. Yet the antics of Rev. May Pepper and her comrades defined the subject for a far wider audience. It was all too easy to see the supernatural as a circus exhibit rather than a topic for serious investigation.

JAMES HYSLOP decided that a bolder approach was required to improve the image of psychical inquiry. He admired what the American branch of the SPR—and Richard Hodgson, in particular—had accomplished, but Hyslop, on his own, conceived of a program that would cleverly combine psychical research with traditional studies. He wanted to encourage scientists to see connections that seemed apparent to him—links such as those between abnormal psychology and medium abilities, hypnosis and trance states. Flournoy's study of Helene Smith certainly provided an illustration of such intersecting fields of study. Hyslop wanted to create a research center that would accommodate both psychical and psychological studies. Once he'd settled on a suitably neutral name for his venture—the American Institute for Scientific Research—he got busy promoting it.

Hyslop wrote to the *New York Times,* describing his new institute as one that could easily become a "research organization of a national character," and revealing that he hoped to raise an endowment of $1 million for its support. He bolstered his claim of national potential by boasting that some of the country's best researchers already endorsed the plan and that the board of trustees already included such eminent men as Professor William James of Harvard. "It is certainly high time that this field should receive the attention of the scientific world in some other manner than mere recognition," Hyslop continued. "The scandal of science is that it has not been endowed as many less worthy causes have been."

William James admired Hyslop's courage and dedication, but found his new associate deplorably lacking in the necessary social skills and graces. Hyslop had "all the heroic qualities of human nature and none of the indispensable ones," James once complained. As a case in point, Hyslop had given James no warning that his name would be used for public fund-raising, leaving the Harvard professor to discover his new role via the newspapers. The result was a stiff letter in which James asked to be excused from the institute's board of directors. He assured Hyslop that the plan was admirable, but "I didn't at all foresee the newspaper campaign and I have enough to carry in the way of reputation for crankiness without shouldering that."

Hyslop had taken James's wholehearted support for granted, never dreaming that James would deny so worthy an endeavor. "I have had a hard enough task to fight this battle," he wrote back. "I ought to find some moral courage in those who have spoken on this subject as you have done. I should expect you to assist in the task." Hyslop had heard from one pledged donor that James had been telling people of his refusal; the woman had promptly withdrawn her pledge. "It is worse still than that, just, when I am at the point of success, failure should be traceable to an act of yours."

James recognized some justice in the accusation. It was reminiscent of what Fred Myers said years earlier, that James failed to lead because he deliberately pulled back from full engagement. He wrote to reassure Hyslop that, even though he was withdrawing from the board, they were indeed on the same side: "My Excellent Hyslop—I am sorry that the rumor of my leaving the society you are founding should cause any donation to cease. I thoroughly believe in endowing research in the psychical direction and had I money to spare I would make over to you several thousands.

"Please show this letter to the friend in question [and previous correspondence if it will help] . . . I repeat my position. I believe in psychical research and its endowment. I disbelieve in . . . the substitution of 'audiences' for investigators, 'popular interest' for investigation and newspaper tattle for facts."

THE NIGHT HELD too many shadows; Leonora Piper could not summon sleep. It was December 20, 1905, in the last bright days before Christmas. She'd gone to bed early, tired by some shopping chores. There was no reason why she should feel so chased by darkness.

She tossed, turned, got up about midnight to make warm milk, returned to bed, but found herself listening for footsteps, troubled by a sense of someone walking around her room. Finally about 1:00 a.m. she fell asleep, only to wake three hours later, still half caught in a dream.

She'd been trying to enter a tunnel, walking toward its entrance. As she approached, she saw a man ahead of her. She could tell that he was bearded, but a slouch hat hid most of his face. She got closer; he raised a hand to block her way. As his fingers reached toward her, she startled awake.

The room was black in the early morning. She went to the window. The casement stood partly open; the rain was beating in. She closed it and went back to bed, tumbling into dreamless sleep. At half past seven, her daughters came into her room. She blinked awake, and the memory of the dream came back in such detail that she immediately began telling them of it.

The hand had looked remarkably like Richard Hodgson's, she told Alta, with its strong long fingers and callused palms. She wished she'd stayed in the dream a little longer, just to be sure, but she was quite certain. It made no sense to her, the hand raised in denial, the shadowed face, the tunnel stretching away into darkness.

Exhausted and still oddly troubled, she decided to stay in bed. An hour later, she heard Alta's feet scrambling up the stairs. Her daughter came in crying, carrying the morning paper's news, which told of the death of Richard Hodgson.

He had gone to one of his favorite hangouts, the Tavern Club, for lunch before a game of handball. An increasingly loud argument occupied the clubroom just then, with one man defending an unpopular cause against angry disagreement.

Hodgson leaned over the stair railing, jokingly emphasizing the unfair, one-sided nature of the fight. "Go for the scoundrel," he boomed. "Don't give him a chance to speak! Down with him, don't let him be heard."

The quarrel broke up in laughter.

"How can anyone be heard when you're in the room, Dick?" countered a friend.

Hodgson had continued on to the Union Boat Club for his regular game. It had barely begun before he'd collapsed on the court, dead of a massive heart attack.

"Absolutely sudden, dropt dead while playing violent handball. . . . All his work unfinished," James lamented. "No one can ever learn those records as he knew them—he would have written certainly 2 or 3 solid books. Too bad, too bad!" That was only part of it though. James mourned more than the loss of yet another of psychical research's best workers; he mourned the loss of "the manliest, unworldliest, kindliest of human beings. May he still be energizing somewhere."

Hodgson's funeral was held three days after his death. The ceremony took place at his beloved Tavern Club. His coffin was decked in ivy, violets, and white roses. Flowers were heaped around the room. After the formal service, his friends gathered round and sang the club song.

James marveled that the bachelor Hodgson, with no family nearby, no group of office colleagues or co-workers, could draw such a crowd: "Everyone was there from simple personal affection for the man in the coffin. I stood at the foot of the stairs and saw everyone come down. *All* the women, and many of the men, were crying."

The rain that had begun earlier that week was still beating down, gray and cold against the windows.

IN THE MIDST of a written message from her trance personality "Rector," Mrs. Piper's pencil dropped onto the paper. Her fingers trembled convulsively, clutching whitely around the pencil when it was returned to her.

"What is the matter?" the sitter asked.

Her hand, still shaking, wrote the letter "H" on the paper, pressing so hard that the point broke. It then continued the word, wrote "Hodgson."

"God bless you!" exclaimed the sitter.

"I am . . ." and then the writing tailed away into wild scrawls.

"Is this my friend?"

The most dictatorial of Mrs. Piper's trance personalities, Rector intervened: "Peace, friends, he is here, it was he but he could not remain, he was so choked. He is doing all in his power to return."

A few days later, the H spirit flickered back again: "I am Hodgson. . . . I heard your call—I know you," he wrote to a young woman sitting with Mrs. Piper.

"Piper instrument. I am happy exceedingly difficult to come very. I understand why Myers came seldom. I must leave. I cannot stay. I cannot remain today."

And then, another two weeks later, on January 23, Alice James and her son Billy came for a sitting. "Why, there's Billy! Is that Mrs. James and Billy? God bless you! I have found my way, I am here, have patience with me. All is well with me. Don't miss me. Where's William? Give him my best wishes."

JAMES CONTINUED TO mourn Hodgson's loss—and the dearth of people to step into his job. Increasingly, he feared for the survival of the small American branch of the SPR. Hodgson had been the only "real worker" in the organization, and the one who best understood the evidence they'd amassed (including twelve boxes of documented Piper sittings read by no one but Hodgson). James could think of no one suited to pick up that monumental workload.

It was true that James Hyslop—"so good but so impossible"—offered to assume Hodgson's duties. But that simply would not do, James told the applicant, listing his defects without ceremony: "You lack the discretion that was so extraordinary a gift of Hodgson's; you are too impulsive; and your enthusiasm leads you to relations with newspapers and the general public that would quickly undo the credit that the name SPR has slowly earned for caution and criticality."

Yet James agreed with Hyslop that psychical research should continue. And as the "Hodgson-control" began to flicker in and out of Mrs. Piper's sittings, it became obvious that here was another opportunity to prove or disprove a returning spirit, much as Hodgson himself had attempted with the G.P. personality.

If he was going to set the standards so high, despite the troubling lack of depth among psychical researchers, if he was determined that the Hodgson-control be studied with the restraint and precision that he demanded, the best investigator—perhaps the only investigator really available—was William James himself.

MARGARET VERRALL HELD a position as classics lecturer at Newnham College. She was fluent in both Greek and Latin, the wife of a Cambridge philosophy professor, an old friend of the Sidgwicks and Myerses, and, most of all, a woman of tireless patience.

Since Myers's death, she'd been pondering his wish to prove immortality. She'd liked him so much, and he'd believed so passionately in survival of the soul, that she wanted someone to make a concerted effort to contact his spirit, if it existed. Finally, as several years passed, as 1905 wound away, she decided to do it herself. As she explained to all those who considered this a strange decision, she felt she'd be letting down a friend if she didn't at least try. And then, too, after all that time spent in the company of Myers and the Sidgwicks, she was a little curious.

An organized woman by nature, Mrs. Verrall worked out a careful system for contacting a spirit. First, she chose automatic writing as the best way to converse with Myers. Then she set aside a time in the late afternoon to try to acquire this skill. Then she waited. For three months she sat at her desk daily, holding a pencil against a piece of paper for at least an hour, listening to the mantel clock tick away the time. Day after day, she arose stiff with sitting, the blank tablet mocking her.

Gradually she became so bored that she quit focusing on the elusive Myers and fell into musing on her work, her garden, her household duties, her family. Lost in that daydreaming haze, she found herself suddenly snapping to attention, the tablet covered with simplistic messages in Greek and Latin—much cruder versions than she usually used—but with the signature "Myers" at the end.

The scribblings seemed almost meaningless, except for one curious coincidence. Over in Boston, Mrs. Piper's spirit guide Rector suddenly began reporting conversations with Myers. And on those particular days, the messages, although in English, were often startlingly similar in content to the Greek and Latin notes taken by Mrs. Verrall. Meanwhile, Mrs. Verrall's daughter, Helen, became intrigued enough to experiment with automatic writing herself. She too discovered that occasionally she'd jotted down a message that was duplicated in notes taken down by her mother or Mrs. Piper.

By early 1906, the three women's writings seemed to form an unlikely kind of chain letter from the dead. Other lost colleagues appeared. Some notes purported to be from Edmund Gurney, some from Henry Sidgwick. Back and forth across the Atlantic, Nora Sidgwick, Oliver Lodge, and William James began comparing the messages. Taken separately, each woman's writing seemed a kind of stream-of-consciousness jumble of words and thoughts. Taken together, though, the messages seemed connected, as if ideas were relayed on some circuit impossible to detect. As she admitted to her friends, Nora began to wonder for the first time if her Henry had been wrong, if there was a chance, after all, of proving conversations between the living and the dead.

IN JUNE 1906, Mrs. Piper received a friendly invitation from Oliver Lodge's wife, Mary. The SPR wanted her to return to England for a second round of investigations—and, more personally, Lady Lodge would be delighted to see her again.

Many strategy discussions and letters had preceded this invitation. Nora, Lord Rayleigh, Oliver Lodge, William James, and a new SPR administrative secretary, a slight, dark Cambridge graduate named John Piddington, all debated how to properly study this curious messaging system.

The invitation came only after the experimental plan was in place. Nora would oversee the Verralls in Cambridge, and Piddington would keep Mrs. Piper sequestered in London. None of the details of the study would be revealed to the three women, only that they were participants in a new series of tests.

The British Society for Psychical Research, thanks to the determination of Nora Sidgwick and Oliver Lodge, had rebuilt itself with some real success. John Piddington was one of two honorary secretaries; the other position belonged to the Hon. Everard Feilding, a younger son of the earl of Denbigh. At the moment, Feilding—who had a known affection for the more peculiar phenomena—was investigating a candle-throwing poltergeist. The more staid and methodical Piddington seemed a logical choice for the correspondence study.

Piddington had a businesslike style about him and a fondness for organization. He had helped set up an endowment for the SPR so that it could pay full-time researchers; he'd managed the transfer of Hodgson's voluminous records to England. To the great appreciation of James Hyslop, he'd transferred some of Hodgson's responsibilities to Hyslop's New York institute, effectively merging the American organizations.

But the cross-correspondence study, as it came to be known, was to be managed strictly by the British SPR. Under that plan, Mrs. Piper would come to England for a series of experiments that began with a rather obvious difference between the two mediums. Mrs. Verrall was a scholar, trained in Greek and Latin. Mrs. Piper had a New Hampshire primary school education and no knowledge of the classic languages. But—and this was the key—the "spirits" in question belonged to men who did know those languages. So if Gurney, Sidgwick, or Myers were actually communicating with Mrs. Piper, they would understand Latin and Greek instructions even if she didn't.

Following that logic, the tests would be conducted in the following manner: Piddington would wait till Mrs. Piper was entranced. He would then ask her, or her control, to give a message to Myers. Piddington would then read off a message in Latin, concluding with a request to relay its content to Mrs. Verrall.

If the message arrived, if they transcended the language barrier, it would be hard to avoid a conclusion that some intelligence greater than that of the mediums was working with them.

By December 1906, Mrs. Piper and her daughters were settled in London, and once the American medium and Piddington had learned to be comfortable with each other, the work began in earnest.

In mid-December, during several sittings, Piddington talked to Mrs. Piper's Rector, asking him to pass along instructions to Myers and his friends. The instructions were given in Latin, each word pronounced slowly, syllable by syllable, and then spelled out for purposes of clarity.

Piddington's message began with a compliment: "Diversis internuntiis quod invicem inter se respondentia jamjudun committis, id nec fallit nos consilium, et vehementer probamus." (As to the fact that for some long time you have been entrusting to different messengers things, which correspond mutually between themselves, we have observed your design, and we cordially approve it.)

This polite opening was followed by a request, also in Latin: Could the Myers personality, once contacted through Mrs. Piper, send a signal to another medium (in this case, Mrs. Verrall)? And could he attach to that message a recognition device, some code words or symbols of his choice?

After several sittings, Rector replied to Piddington: "We have in part understood and conveyed your message to your friend Myers and he is delighted to receive it so far as he has been able to receive it." Several weeks later, in early January 1907, Rector seemed to think that Piddington might need some reassurance as to the delay: "Hodgson is helping Myers with his translation."

Two weeks later, Rector said he had a communication from Myers: "I should like to go over the first and second sentences of our Latin message. . . . I believe I can send you a message which will please you if I understand it clearly."

By this time, Piddington had given further thought to the idea of a recognition device. He had a more specific suggestion, again conveyed in Latin: Could Myers please ask the medium at the other end to draw a circle and triangle as part of her response?

That night, Mrs. Verrall wrote: "Justice hold the scales. That gives the words but an anagram would be better. Tell him that—rats, star, tars and so on. Try this. It has been tried before. RTATS. Rearrange these five letters or again t-e-a-r-s . . . s-t-a-r-e."

Five days later, she wrote: "Aster [a star] . . . the world's wonder, And all a wonder and a wild desire/the very wings of her. . . . but it is all much the same thing—the winged desire, the hope that leaves the earth for the sky . . . Abt Vogler for earth, too hard that found itself or lost itself—in the sky. That is what I want, On the broken sounds, threads."

She closed the message with a circle and a triangle.

On February 11, Rector delivered another message from Myers, one that gave some clues as to what he was trying to do. He told Piddington that *hope, star,* and Browning were all important in Mrs. Verrall's script.

With that, Nora Sidgwick realized that all those ramblings about stars made actual sense. Myers was ever a lover of poetry—and "Abt Vogler" was a poem written by Robert Browning. It was a tale of a musician, included in that same 1864 book that featured "Mr. Sludge, the Medium."

Nora hurried to find it among her poetry books:

And the emulous heaven yearned down, made effort to reach the earth,
As the earth had done her best, in my passion, to scale the sky:
Novel splendours burst forth, grew familiar and dwelt with mine,
Not a point nor peak but found and fixed its wandering star.

"The mystic three," wrote Miss Verrall on February 17, "and a star above it all / rats everywhere in Hamelin town / now do you understand?" She had been drawing as well—a crescent moon, a star, and a winged bird. Her illustrated message was signed "Henry."

Mrs. Verrall now wrote down a message, signed from Myers, saying that he was worried that Rector did not know the poem so familiar to the rest of them: "I am most anxious to make Rector understand about the name of that poem."

Some weeks later, while entranced, Mrs. Piper carefully wrote the words, "Abt Vogler."

"Now, dear Mrs. Sidgwick, in future have no doubt or fear of so-called death, as there is none, as there is certainly intelligent life beyond it."

Mrs. Verrall was writing messages, purporting again to be from Myers. "Yes, it's a great comfort," Nora replied.

"Yes, and I have helped proclaim it for you all," the Myers script continued, explaining that he had chosen the Browning poem because it best fitted his own life, wandering the stars. He had more to say, but it was so incredibly frustrating getting even the smallest shred of a thought across. Myers hadn't realized in life how difficult it would be—even between old friends—to reach through the drawn curtains of death.

"You must patch things together as best you can. Remember we do not give odd or singular words without a deep and hidden meaning."

MEANWHILE, Piddington found himself fixated on the sequence of recurring words in the messages, the repetition of star, rats, arts, stare. They reminded him of anagrams, which had been one of Richard Hodgson's favorite recreations. Hadn't Mrs. Piper written earlier that Hodgson was helping with the translations? And hadn't Mrs. Verrall made a reference to anagrams?

Piddington wrote to ask William James if there were any papers in Hodgson's effects, those still remaining in Boston, that contained anagrams. James's oldest son, Harry, who was co-executor with his father of Hodgson's estate, obligingly sifted through the personal papers remaining.

In one of the boxes, Harry James found a sheet covered with Hodgson's cramped scrawl, which he mailed to London. It was a practice for an anagram, and it read:

RATES
STARE
TEARS
TEARS
TARE

ARE ST
ST ARE
A REST
REST A

STAR
TARS
RATS
ARTS
TRAS

"I confess that when this came into my hands I felt as I suppose people do who have seen a ghost," Piddington said, shaken out of his usual calm by that creased sheet of notepaper.

REV. MAY PEPPER continued her career as a star of the New York media and the bane of James Hyslop's plans for restoring scientific credibility to psychical research.

Traditional psychologists had shown no interest in Hyslop's institute, scuttling his plans to integrate their work with his less orthodox interests. But thanks to John Piddington's intervention, he'd gained sanction to rename his research institute the American Society for Psychical Research and build on the old organization's foundation.

Hyslop had barely started a new round of fund raising and assurances that psychical research was a meaningful science when Pepper went on trial for using her "spiritualistic gifts" to bilk a wealthy industrialist out of home and fortune.

Since her temple days, Mrs. Pepper had married an ardent follower, Edward Ward Vanderbilt, a widower who'd made a fortune in the wholesale lumber business. She'd resigned her pastorate and moved into a large house in Brooklyn, "one of the handsomest in the Bedford district," according to the *New York Times*.

The reverend, her elderly husband, and her spirit control, an Indian child she called "Little Bright Eyes," planned a simple and happy life together, she told reporters—a life enlivened by occasional private séances. But not if Vanderbilt's daughter could prevent it; she had demanded a formal hearing into her father's sanity.

At the hearing, witnesses revealed that Mr. Vanderbilt had given Little Bright Eyes gifts of candy, clothes, and cash, illustrating, as the *Times* gleefully noted, that "girls in Spookland, as here, have a weakness for bon-bons and caramels at 80 cents a box, that they can find use for wearing apparel and can cash checks."

The spirit also exhibited a talent for giving very specific instructions to Mr. Vanderbilt. Letters produced in court revealed that through Bright Eyes, his dead wife had told Vanderbilt that she wanted the widower to marry May Pepper and, of course, to provide her with a comfortable home. "Bright Eyes is so good to me," ran one such note, "that it behooves me to do all in my power to make her Medy's [medium's] life brighter and happier."

The letters were delivered through automatic writing by May Pepper herself, who insisted that she never knew what her hand wrote while she was lost in a trance. Any wrongdoing, Mrs. Pepper-Vanderbilt told the court, was due to the spirit's mischievous nature. A medium could not be held responsible for the morals of the ghosts that possessed her.

When the case had been fully presented, the jury agreed unanimously with Vanderbilt's daughter that her father had lost his mind. No one could miss the corollary—that only lunatics wasted their time believing in ghosts and well-meaning spirits.

As Hyslop saw it, his only choice was to fight fire with fire, to show that his organization could expose frauds with the best of them. As May Pepper needed no more exposure—or publicity—he selected another target, equally high-profile in its way.

First, Hyslop hired a full-time investigator for the ASPR. His choice was Hereward Carrington, a soft-spoken young Englishman with a reputation as both an amateur conjurer and psychical skeptic. Carrington's first assignment was to visit the western New York community of Lily Dale, billed as "the most famous and aristocratic spiritualistic camp in America."

A mere scrap of a town about sixty miles south of Buffalo, Lily Dale contained little more than a railroad station, a few hotels, a scatter of white-painted houses, and an assembly ground. Nothing out of the ordinary except that the Church of Spiritualism owned it all. Only mediums registered to the church could become full-time residents of Lily Dale. Believers said that such a concentration of spiritualist power drew ghosts to the town the way a magnet summoned iron filings.

The hotels, the mediums' parlors, and the wooded grounds—supposedly haunted—were open to any paying customer seeking advice, communication with the dead, or just a demonstration of unearthly powers. During the summer months, Lily Dale hosted some 500 visitors daily, all hoping to catch a glimpse of the ghost world. As Hyslop put it, "Such places simply invite investigation by the claims that they make."

Since he was already a known critic, Carrington registered at his Lily Dale hotel under the name Charles Henderson. To his surprise and amusement, he discovered that the mediums of Lily Dale specialized in old-fashioned performances—slate writing, cabinets, and materializing spirits—which told him to set his expectations very low.

In short order, Carrington caught one medium busily switching slates, another smuggling confederates into a cabinet. His favorite moment—as reported in the *New York Times*—occurred when the lights accidentally came on during one cabinet séance, and a young female spirit, "instead of melting away," darted panic-stricken, gauzy skirts flying, back into the cabinet.

Hyslop's complaint was that the newspapers seemed more fascinated by the scams than by the excellent investigation. The *Times* headlined Carrington's report "Ingenious Frauds at Lily Dale Séances." He wished it had been presented as "Brilliant Investigation by Psychical Researchers."

PORING OVER THE cross-correspondence scripts, Nora's assistant at the SPR, a Newnham College graduate named Alice Johnson, suddenly remembered a peculiar letter received from India a few months earlier.

The unexpected correspondence came from Alice Kipling Fleming, a sister of writer Rudyard Kipling and a longtime secretive psychic. For years, Mrs. Fleming, the wife of a British army officer, had been troubled by an uneasy sense of the occult. She did her best to keep her feelings secret, though, because her family disliked the subject.

"It puzzles me a little," Mrs. Fleming wrote to Miss Johnson, "that with no desire to consider myself exceptional I do sometimes see, hear, feel or otherwise become conscious of beings and influences that are not patent to all. Is this a frame of mind to be checked, permitted or encouraged? I should like so much to know. My own people hate what they call 'uncanniness' and I am obliged to hide from them the keen interest I cannot help feeling in psychic matters."

Mrs. Fleming had read Myers's book *Human Personality and its Survival of Bodily Death,* and it had inspired her to begin surreptitiously experimenting. After a recent afternoon of automatic writing, she'd found some lines that concluded with the signature "Myers." They included some remarkably specific direction to send the text to Mrs. Verrall in Cambridge.

Mrs. Fleming didn't know Mrs. Verrall, didn't know if the directions were real, doubted whether any part of the message was authentic—but it was precise enough to make her feel that she should do something. She decided to hand off her scripts to the SPR: "Will you forgive me for troubling you with the writing? I do not like to suppress it as it gave me the impression of someone very anxious to establish communication, but with not much power to do it . . ." If they did choose to use her work, Alice Fleming asked Miss Johnson and her friends to protect her by using a false name. They did. In SPR publications, she was known only as Mrs. Holland.

When she took a second look at the scripts from India, Miss Johnson found that the so-called Myers had given Mrs. Fleming a near perfect description of rooms in Mrs. Verrall's house. But even more curiously, she'd written other details that suggested that Mrs. Fleming had unwittingly been pulled into their cross-correspondence experiments.

ON APRIL 17, 1907, Mrs. Piper suddenly began fumbling for the Greek word for death. "Sanatos," she wrote, haltingly. "Tanatos." Then several days later, it came out right: "Thanatos, thanatos, thanatos."

Death, death, death.

One day earlier, Mrs. Fleming had mailed a script from India that read in part, "Maurice, Morris, Mors." The last was Latin for death; it seemed to Miss Johnson that their India correspondent was reaching for the counterpart to Mrs. Piper's thrice-times death. And Mrs. Fleming continued, "And with that the shadow of death fell upon him and his soul departed out of his limbs."

A week later, Mrs. Verrall wrote, "Pallida mors" (Pale death), and then, "Warmed both hands before the fire of life. It fails and I am ready to depart."

. . .

THE CROSS-CORRESPONDENCE experiments filled hundreds of pages. Not all connected so neatly; not all even made sense. But enough did; enough of those flares of similarity brightened the pages that the investigators saw only two meaningful choices. They must either accept a pattern of exceptional coincidences or accept that they were reading mental messages sent and received by both the living and the dead.

Almost all the psychical researchers reached the latter conclusion. They wished, of course, that the spirits could do a better job of getting their precise message across, that the results would be more exact, that the proof would be more convincing to their critics. They were told, as the correspondence continued, that the spirits wished, in turn, that their human contacts would do a better job as well.

"Back in the old despondency," read one passage, taken down by Alice Fleming and signed "Edmund Gurney." "Why don't you write daily? You seem to form habits only to break them."

Mrs. Fleming told Alice Johnson that the complaint spilled out after she had been too busy to spare time for automatic writing. "If you don't care to try every day for a short time, better drop it all together. It's like making appointments and not keeping them," the Gurney message continued. "You endanger your own powers of sensitiveness and annoy us bitterly."

Some of the messages signed by Myers seethed with frustration: "Yet another attempt to run the blockade—to strive to get a message through—how can I make your hand docile enough—how can I convince them?

"The nearest simile I can find to express the difficulties of sending a message is that I appear to be standing behind a sheet of frosted glass—which blurs sight and deadens sound—dictating feebly—to a reluctant and somewhat obtuse secretary.

"A terrible feeling of impotence burdens me."

WILLIAM JAMES WAS a retired professor now, having taught his last class at Harvard in January 1907. At the age of sixty-five, he seemed noticeably thinner and grayer, but he assured his many well wishers that he planned only to slow down a little, to put more time into his work as a philosopher.

Of course, James also spent many of his newly liberated hours in the company of the self-proclaimed spirit of Richard Hodgson. His desk stood stacked with piles of transcripts, records of the sittings held before Mrs. Piper left for England to do the cross-correspondence work.

He was sifting, analyzing, fuming: "It means much more labor than one would suppose, and very little result," he wrote to his brother, Henry. "I wish that I had never undertaken it."

Encounters with the Hodgson control veered between a presence so real that James remembered breaking out in a chill during the sitting and, at the other extreme, tedious hours with what appeared to be some peculiar creation derived from Mrs. Piper's interpretation of the masculine personality.

The Hodgson control tended to announce himself with the unfamiliar heartiness of a campaigning politician, exclaiming, "Well, well, well! I am Hodgson. Delighted to see you. How is everything? First rate?"

Hodgson had never talked like that in his life.

Yet the glad-handing usually gave way to a familiar friendliness, as if the ghost—if it was such—had to pull free from Mrs. Piper before emerging as himself.

The spirit Hodgson teased his old close friends, turned quiet and serious with those he knew less well. One woman had told James that she and R.H. "were such good friends that he was saucy toward her, and teased her most of the time," which was exactly as the control treated her, "*absolutely characteristic* and as he was in life."

Another former friend left his sitting feeling dizzy and shaken. After the irritating greeting period "came words of kindness which were too intimate and personal to be recorded, but which left me so deeply moved . . . it had seemed as though he had in all reality been there and speaking to me."

James was determined to be as ruthless an investigator as Dick Hodgson had ever been. Emotional responses were all very well, but they weren't facts. And facts didn't count until they'd been dissected into pieces and every fragment examined.

In one sitting, the Hodgson personality had asked a friend to destroy some letters written to a woman and hidden in his desk. "Look for my letters stamped from Chicago. I wouldn't have them get out for the world."

Unable to find the letters, which had apparently been shipped to England already, James had demanded further information about the woman in question. Came the response: "There was a time when I greatly cared for her and I did not wish it known in the ears of others. I think she can corroborate this. I am getting hazy. I must leave."

When he returned, the spirit Hodgson added another detail. The last time he'd seen the woman, "I proposed marriage to her, but she refused me."

James had never heard a word of this from Hodgson when alive. He doubted the claim. Nevertheless, he wrote to the woman in question, receiving an answer that he hadn't expected: "Regarding the utterances of Mrs. Piper, I have no difficulty in telling you the circumstances on which she may have founded her communications. Years ago, Mr. H asked me to marry him, and some letters were exchanged between us which he may have kept."

Holding her letter, James felt a leap of euphoria. In this secretive part of Hodgson's life, it seemed possible to demonstrate "the return of a 'spirit,' proven by extremely private knowledge." The trick, of course, was in the privacy of the knowledge. Carrying through with his resolve to investigate thoroughly, he interviewed a dozen of Hodgson's friends. None knew of the relationship, but one did tell him that Hodgson occasionally consulted with Mrs. Piper's Rector on his private life. So there was a possibility, admittedly slim, that Mrs. Piper knew about the hidden letters and the failed proposal. She showed no signs of knowledge, but perhaps it was locked away in a "trance memory." On that slim possibility—and he knew some would consider it slim to the point of being meaningless—James found he could not quite declare for proof of spirit communication. He admitted, though, that by such standards one might never find acceptable proof. "It would be sad indeed if this undecided verdict will be all I can reach after so many years," he wrote to Flournoy.

WITH JOHN PIDDINGTON consumed by the cross-correspondence experiments, many of the SPR's other endeavors fell to Everard Feilding, a forty-year-old aristocrat with a quick intelligence, an irrepressible sense of humor, a love of odd causes, and—since the drowning death of his twin brother in 1906—a serious interest in the possibility of survival after death.

Feilding had found his Christian faith shaken by the tragedy, the consolations of his church unconvincing. The alternative offered by modern science—German biologist Ernst Haeckel ascribed the notion of soul to "plasma movements in the ganglion cells"—struck him as even less sustaining. With clergy decrying science and scientists denying religion, Feilding had turned to psychical research as the last hope for a man of rational faith. At least, he felt, this was a group trying to make some sense of the world, displaying prejudice against neither flesh nor spirit.

Despite that underlying seriousness of purpose, Feilding sincerely enjoyed the comedy that seemed ever to color psychical research. In shredding the reputation of a physical medium working in London, he wrote, "He promises good conditions, and gives them; so good, in fact, that his fraud is perfectly obvious. It is though having told you that a spirit would appear in full daylight from the next room, I presently were to emerge thence, dressed in a white sheet and a mask, and say I was your grandfather."

Feilding admitted to having a wonderful time questioning the medium's manifested spirits, even pretending to recognize them:

"What! Sidney Parry!"

(It was really quite like him but Sidney Parry isn't dead.)

Great emotion and assent.

A few other words from me, ending with, "But my dear fellow! When did you pass over?"

The spirit, Feilding reported to his colleagues, fled the room. He'd envied the muscular, energetic nature of its stride.

He'd seen nothing to convince him that spirits existed or that mediums could contact them. But the theater was so good, the fraud so entertaining, that Feilding couldn't persuade himself to abandon it.

.   .   .

CHARLES RICHET HAD renewed his request to Nora Sidgwick that the SPR take another look at Eusapia Palladino. Nora kept repeating that she wanted no further contact with the grubby and distasteful Eusapia. The British psychical researchers were busy with the cross-correspondence studies, which Nora thought far more promising than the parlor trickery of levitating tables.

But she liked Richet and admired his persistence. Neither Richard Hodgson nor her husband was there to dissuade her. Perhaps, she thought, it wouldn't hurt to take one more look. And in that case, Nora thought, the irrepressible Everard Feilding was just the man for the job.

Feilding—after a strenuous discussion about investigations that wasted one's time—agreed to visit the notorious Eusapia. But he demanded reinforcements for his trip to Naples: a talented amateur magician named Wortley Baggally and the skeptical investigator Hereward Carrington. The visit was scheduled to begin in late November 1908. None of the three rejoiced in the mission.

Baggally had agreed to go only because Feilding was a friend, but he insisted on announcing in advance that he believed Eusapia to be a fraud. Carrington was depressed by the whole assignment; he had "never seen anything that he was unable to account for by trickery," and he usually finished such assignments convinced that he was a better conjurer than the person under investigation. In turn, Feilding had made it clear to Nora Sidgwick and to Oliver Lodge that he would do the job, but he expected little more than the usual clumsy deceit—and perhaps a few good laughs.

Resentfully, the three investigators planned their campaign. They set up camp in the small Hotel Victoria in Naples, taking adjoining rooms so they could stand watch together. They decided to use Feilding's room for the tests and to prepare it themselves.

First, they hung four bare lightbulbs from the ceiling, all clustered above the chair they had designated for Eusapia. Two of the bulbs burned brilliantly, and the others shed a softer light. The combined effect was of sitting under a small sun.

Eusapia meekly accepted their conditions, making only two requests, that the preparations include a small wooden table and that a pair of curtains be hung across a corner. They could procure them, or she would lend her own. After talking it over, they decided to use her curtains and table. It would give them an early chance to catch her cheating. But the curtains turned out to be thin, nearly transparent black cashmere, and the table a rimless thing of fir wood, about two and a half feet tall, with a plain circular top. They hung the curtains across a corner of the room, to simulate a cabinet.

She asked them to provide some objects that the "spirits" might use in play. Those they bought themselves: a pair of tambourines, a guitar, a toy trumpet, a tin whistle, a toy piano, and a china tea set. Each day, they would hide different objects behind the curtains, changing them without notice.

Feilding admitted that some might criticize these arrangements as too obliging, that they should have insisted on novel effects, séances without her usual curtains and levitating objects. His answer was that they wanted to study Eusapia in standard conditions.

When they were finally ready, two of them stayed in the room to keep guard; another went down to the hotel lobby to meet her, making sure that she entered the test room alone.

Some mornings, Eusapia arrived apparently uninterested. She would rather gossip, rattle away until she "had tired herself out with her own conversation." Eventually she would begin to yawn. This was a favorable symptom, Feilding said, "and when the yawns were followed by enormous and amazing hiccoughs, we knew it was time to look out, as this was the signal for her falling into a state of trance."

The next day she might stomp silently into the room and tumble into trance, complaining of fatigue. On those days, they usually sat with nothing to watch but the minutes ticking away. They kept notes anyway—good day or bad—for eleven sittings, each man recording his own impressions, tracking who was holding her hands, her feet, her knees, who was above the table, who was below it.

The curious thing, Feilding noted, was that they got some of their best effects when they were holding tightest. Not that she didn't often try to wiggle away, especially on her bad days. But if they weren't sure of her, they simply wrote off the results.

On good days, though, they weren't sure what to write off.

With all the lights blazing, with Carrington gripping her hands and Baggally draped across her feet, the little table floated two feet off the ground and hung there. "No precautions we took hindered [the levitations] in the slightest," the investigators reported. "She had no hooks, and we could never discern the slightest movement of her hands and knees. We very often had our free hands on her knees, while her feet were controlled either by our feet or by one of us under the table, and were generally away from the [levitating] table legs, a clear space being discernible between her and the table."

The curtains blew toward them when she stretched out her hand, swirling like smoke: "There was no attachment to her hand as we constantly verified by passing our hands between her and the curtain. Nor would any attachment produce the same effect, as the curtain was so thin that the point of attachment of any string would have been seen."

Lights glimmered overhead: a steady glowing turquoise, a yellow streak, a "small, sparking light like the spark between the poles of a battery." Invisible objects solidified in the air, poking and prodding the researchers. Sometimes they formed into "more or less indescribable objects; white things that looked like handfuls of tow [fibers of flax], stalk-like bodies which extended themselves over our table, shadowy things like faces with large features, as though made of cobweb."

In response to a query from England, asking if they'd seen any of Richet's so-called ectoplasmic forms, Feilding wrote back to Alice Johnson, "Why my good lady, we are getting hands, white & yellowish; heads, profiles & full face; curious black long knobbly things with cauliflowers at the end of them; touches, visible & invisible; hand grasps from within the curtain, one yesterday (12/05/08) which held my hand with such force that I felt the nails."

All three men finished shaken, puzzled—and convinced.

The phenomena in question were "in themselves, preposterous, futile, and lacking in any quality of smallest ethical, moral or spiritual value," Feilding said. Both Baggally and Carrington were capable of conjuring up similar effects on a stage, given adequate props and preparation. The investigators were less impressed by the show itself than by an almost tangible sense of magic filling the room, an unnerving impression of a power beyond any of them, including Eusapia herself.

Feilding did not believe that he'd seen spirits at work, rather that he'd witnessed something less definable, "the existence of some force not yet generally recognized which is able to impress itself on matter and to simulate or create the appearance of matter. Thus the demonstrations—blowing curtains and floating tables—might seem rather silly except as a clue to something more, the other power in the room."

Feilding still believed that psychic investigators needed to "approach the investigation of the phenomena themselves in a light, shall I say, even a flippant spirit. I sometimes think that in this way alone one can preserve one's mental balance in dealing with this kind of subject."

But in the same way that Piddington had looked upon Dick Hodgson's old star anagrams and felt a ghost breathing down his neck, so Feilding found himself sure that he'd seen beyond the ordinary, that Eusapia's phenomena were "the playthings of the agency which they reveal." On a rare note of complete earnestness, he recommended that the study of that agency, whatever it was, "is surely a task as worthy of the most earnest consideration as any problem with which modern science is concerned."

WILLIAM BARRETT HAD been waiting for that conclusion for more than a decade. Since 1895, to be precise.

That was the year that Barrett had written a book on the importance of psychical research, inspired by the moment when Lombroso and Richet had declared in favor of Palladino. Following the disastrous Cambridge sittings, Barrett had quietly tucked the manuscript away: "It seemed wiser . . . to delay the publication of the volume until more conclusive evidence, one way or the other, had been forth-coming." As soon as he learned of the results of the Naples experiment, Barrett retrieved his manuscript and sent it to a publisher.

It occurred to him that the timing was even better now. As author, he could stand on a lifetime of scientific achievement. Barrett had been knighted for his work in physics, shortly after Oliver Lodge achieved that honor, and been named a Fellow of the Royal Society. As Sir William Barrett, he believed his opinions would carry greater credibility.

Barrett's book, *On the Threshold of a New World of Thought,* rang like a victory cry when it appeared in 1909, declaring that he and his colleagues had proved their case, from the newest results with Eusapia Palladino to more than twenty years of consistent evidence provided by Leonora Piper. "The paramount importance of psychical research," Barrett wrote, "is found in correcting the habit of Western thought . . . that the physical plane is the whole of nature, or at any rate the only aspect of the universe which really concerns us."

His book immediately sold out, rolling right into a second edition. At that moment of multiple successes, Barrett had no doubt that public opinion would carry science along with it and that—despite the fact that they were so outnumbered by their critics—the process of "correction" was finally under way.

1 2

⤻

A GHOST STORY

*O*N JANUARY 10, 1909, a headline in the *New York Times* caught perfectly the flicker of hope warming the psychical research community: "Sir Oliver Lodge Gives the Results of a Series of Remarkable Experiments Testing the Reality of Life After Death."

Keeping with William Barrett's theme—that proof was at hand—Lodge focused on the cross-correspondence study, explaining that he and his friends had kept the results to themselves until they were confident of what they meant. Only now would Lodge reveal a startling series of exchanges, pieced together over several months during 1907.

On a January morning, Mrs. Fleming, from her home in Calcutta, had written a script that included the names "Francis and Ignatius." Five hours later in London, taking into account the time difference, Mrs. Piper's penmanship suddenly changed during an automatic writing session, taking on the distinctive loops of Fred Myers's handwriting. John Piddington asked immediately whether this was Myers and whether he was consulting with any other spirits. The prompt reply came in the form of two names scrawled across the paper: "Francis and Ignatius."

A few weeks later, Mrs. Verrall in Cambridge drew three converging arrows. Both she and Nora puzzled over the drawings and decided not to mention them but instead to wait and see if anything related came from London or Calcutta.

The next day, as Piddington worked with Mrs. Piper, she produced a message from Richard Hodgson, saying that as a test, he had given "arrow" to Mrs. Verrall. A few days later, another message, signed Hodgson, reminded Piddington to "watch for arrow."

At week's end, Piddington received a letter from Nora Sidgwick, for the first time telling him of the arrow scripts at her end. He decided not to mention it to Mrs. Piper. The following day came a slightly impatient message from Hodgson, "Got arrow yet?" Piddington waited two months before letting the medium know the answer to that question, wanting to be sure that he didn't influence the results himself.

"Whatever the agency is that effects the coincident phenomena, it is not a force that is working blindly and mechanically, but with intelligence and design," Lodge said, promising more to come as the cross-correspondence tests were still ongoing, and still continuing to produce surprising linkages.

"Not easily or early do we make this admission," he emphasized, but Lodge was now willing to commit himself publicly, to say that he was communicating with the spirits of his old friends, that on the other side, the messages were sent "with the express purpose of patently proving their known personalities," that it would be soon impossible to deny such proofs of immortality.

William James's mood was chillier. He was writing up his analysis of Mrs. Piper's Hodgson personality into a formal report for the SPR. The self-proclaimed spirit still seemed troublingly ambiguous to him, and he worried that his uncertainties might dampen the feeling of confidence rising from the cross-correspondence work.

James's health was faltering again, draining him of optimism and energy. While resting—relaxing at Chocorua, reading in his Cambridge home library—he felt comfortable enough. But with even mild exertion his heart clutched painfully in his chest.

He barely stirred from home while he struggled to make sense of what had been real, and what had not, in the shadowy reappearances of Richard Hodgson.

The Hodgson spirit or personality or control, whatever one wanted to call it, seemed sometimes a believable ghost, other times an uneven re-creation shaped by the medium's mind and memory. But when believable, it could be downright eerie.

William Newbold, a professor from the University of Pennsylvania and a longtime friend of both James and Hodgson, had come to sit twice with Mrs. Piper, maintaining an attitude of easy skepticism with some difficulty.

"I heard you and William discussing me. I stood not one inch behind you," the Hodgson-control said during Newbold's second visit.

Sitter: William who?

Hodgson personality: James.

Sitter: What did William James say?

Hodgson p.: He said he was baffled, but he felt I was talking at one moment, and then at another he did not know what to think."

(Perfectly true of my conversation with Newbold after his sitting with Mrs. Piper a week previous—WJ)

Sitter: Did you hear anything else?

Hodgson p.: Yes, he said I was very secretive and careful.

"Did you hear him say that?"

"He did. He said I was [keeping information back]."

"I don't remember him saying so."

"I tell you, Billy, he said so."

Newbold related this exchange to James, leading to a discussion of the conversation in question. The Pennsylvania psychologist remembered that James had expressed puzzlement over the on-off nature of the Hodgson personality. He didn't recall any mention of Hodgson's secretive nature.

But James did.

He'd told Newbold that sometimes the Hodgson personality seemed to be protective of what he knew, refusing to answer questions, in a way that brought back the living Hodgson's obsessive demands for secrecy.

That neurotic sense of privacy had colored the sittings, making it more difficult for James to verify information. But Hodgson had been a complicated personality in his lifetime; should anyone have expected him to become an easy man after death?

The impression of a strong-willed and opinionated Richard Hodgson—again, as he was in life—resounded during a sitting with James's wife, Alice. She'd asked, "Do you remember what happened in our library one night when you were arguing with Margie [her sister]?"

Alice James had hardly finished the sentence before the medium's arm shot forcefully out, one hand forming a fist, shaking the clenched fingers angrily into the air. "Yes, I did this in her face. I couldn't help it. She was so impossible to move. It was wrong of me, but I couldn't help it."

James and his family had laughed over that incident. The explosive response had come during a family dinner, when Hodgson and Margie Gibbs were both invited, and James's sister-in-law gave an admiring description of a slate writer she had seen in California. Hodgson—who never could abide slate writers—had become so exasperated that he'd leapt loudly to his feet to challenge her, his fist waving under her nose.

The private jokes, intimate details, embarrassing recollections that the sitters asked James not to mention—in an era not yet much removed from the buttoned-up repression of the Victorian age—left a powerful impression. "More than this—most of us felt during the sittings that we were, in some way, more or less remote, conversing with a real . . . Hodgson," James wrote in his final report on the Hodgson control.

Reading the transcripts of a sitting was always a poor substitute for attending one. James had heard many scientists express doubts about séances after they had read an investigator's report. He felt that without personal experience it was difficult, if not impossible, to understand the event. He could try to describe it, that overwhelming sense of intimacy "when your questions are answered, and your allusions understood; when allusions are made that you understand, and your own thoughts are met . . . when you have approved, applauded or exchanged banter, or thankfully listened to advice you believe in, it is difficult not to take away an impression of having encountered something sincere in the way of a social phenomenon."

But put plainly, the best writer in the world could not convey that shock of personal recognition.

Even so, and for all those extraordinary moments of connection, there were also extraordinary disconnects, ones that seemed to James anyway, to dispel any certain conclusion. The Hodgson personality—simply called R.H. in James's report—could not accurately describe his own childhood in Australia. Neither could he accurately describe his personal life. When asked to name some of his friends at the Tavern Club, men that he had played pool with, gone swimming with, R.H. gave six names, only one of whom belonged to the Tavern Club, and never showed any awareness of that error.

The best results came from the trance personality's knowledge of relationships and experiences that Hodgson had shared with people sitting in the room—making telepathy a better answer than spirit communication, suggesting that the medium might be picking up information from her visitors. And, James emphasized, even these were not the best ever sittings with Mrs. Piper. Along with R.H. came "so much repetition, hesitation, irrelevance, unintelligibility, so much obvious groping and fishing and plausible covering up of false tracks . . . the stream of veridicality [truth] that runs through the whole gets lost as it were in a marsh of feebleness."

In Mrs. Piper's trances, James had often seen hints of the supernatural, the blown spark of unearthly powers glinting beyond his reach—but that was less so with R.H. The Hodgson personality, James thought, was not as strong as the irascible Phinuit, the serious-minded G.P., or even the dictatorial Rector. James still believed that some force, some power, was attempting to communicate in these sittings. But, he concluded unhappily, "if asked if the will to communicate be Hodgson's or be some mere spirit-counterfeit of Hodgson, I remain uncertain and await more fact."

.   .   .

NORA SIDGWICK WAS as fragile in appearance and as tough in mind as ever. As the newly named SPR president, she ruffled her colleagues by agreeing with William James that psychical researchers still must travel over "considerable ground" before they could assume they were conversing with spirits.

She believed that the society had made a powerful case for telepathy—in Mrs. Piper's sittings, in observations, and in other experiments, possibly including the cross-correspondence work. As Nora put it, "few people who looked into the evidence we [already] have, thoroughly and without prejudice, would fail to be convinced that telepathy is a fact." The difficulty was, as ever, to overcome prejudice, especially among scientists.

"We must not altogether blame men of science for feeling this prejudice. It is a kind of self-defense," Nora said. As she saw it, science relied on a carefully defined and validated set of guidelines, which, among other advantages, prevented researchers from chasing wild ideas down dark and peculiar alleys. Frankly, Nora admired such practical efficiency. She reminded her colleagues that the scientific process worked incredibly well, beyond most people's wildest expectations.

In that very year of 1909, German bacteriologist Paul Ehrlich devised the first modern chemotherapeutic drug, a breakthrough treatment for syphilis; wireless communication—which had won Marconi and Ferdinand Braun (but not Lodge) the Nobel Prize—was so well established that it now served as the primary communication method for oceangoing ships; American Leo Hendrik Baekeland unveiled Bakelite, a plastic resin that did not soften when heated, launching the modern plastics industry; a small biplane had been flown across the English Channel; and the success of moving picture technology had inspired producers to build "studios" in a tiny southern California community called Hollywood.

Science proved its power and worth every day, Nora said, and for many educated people it had replaced religion as the most believable way to explain the world. Yet in any belief system, she pointed out, there was a risk of blindness, especially when it became unquestioning belief. "Danger only arises when the scheme becomes a system of dogma which is master instead of slave," Nora said.

Consider, for instance, the insistence of scientists that no observation or experimental result is proven unless it can be reliably repeated. Obviously, there were exceptions to that in nature—no one claimed that a shooting star could be replicated. And yet researchers used their set methods to deny everything the SPR wanted to study, which was also, by nature, spontaneous, erratic, and unpredictable—including the telepathy results.

Under identical conditions, the SPR investigators achieved apparently perfect mental coordination on one occasion, and absolutely nothing on another. "This is one of the difficulties which make patience and perseverance such essential qualities in psychical research," Nora acknowledged, "and it is one of the difficulties which we hope further study may reduce."

As an illustration, she gave the history of a small experiment she'd tracked over the past several years. Two women, who lived in different villages some twenty miles apart, had been daily noting certain impressions or events, trying to send them telepathically to the other, and also mailing postcards recording what had been sent and what had been received.

The postcards illustrated a typical hodgepodge of success and failure. Participant A had attended a tea at which a village woman had worn the oddest pair of spectacles. She decided to send the ideas of the spectacles that day. She received a postcard from participant B complaining that she'd spent the day thinking about spectacles.

On a bright fall day, participant B had been told a lively story by a friend, a tale of a big white hog with an unusually long snout. Amused, she sent a card in the afternoon mail telling of the pig. Participant A sent a postcard that same day, which talked of a cold, wet evening and a pig with a long snout.

"The setting you see was wrong," Nora noted, "but the pig turned up all right." She doubted that a traditional scientist would see the pig and the spectacles as evidence. But she did. She regretted that scientific "dogma" blinded intelligent people to such possibilities.

G. STANLEY HALL, president of Clark University, an ASPR dropout and for many years an outspoken critic of psychical research, had surprised Mrs. Piper's supporters by asking for sittings with her. Hall offered assurances that his intentions were only to do good science. And yet William James did not quite believe him.

Hall had written to both Hyslop and James, asking help in scheduling time with the medium. James inclined toward denying the request. He considered Hall "a crank," a rigid and unforgiving man, and a somewhat vindictive scientist. Hyslop countered that psychical research would never advance by denying mainstream scientists access to its best study subjects.

While the Americans hesitated, Hall continued to press his case. He wrote to Oliver Lodge, asking whether the English SPR "would raise any objection to his making some simple physiological tests on Mrs. Piper's condition in trance." Lodge also thought this looked like an opportunity. Hall, as an eminent psychologist, founding editor of the *American Journal of Psychology*, and mentor to James McKeen Cattell and Joseph Jastrow, was capable of winning the field back to their side.

Still, Lodge wrote, it worried him that James, who knew Hall better, remained so unenthusiastic. Usually decided in his opinions, Lodge opted to go along with whatever James decided was best.

IN THE WAKE of the SPR's cross-correspondence study, especially in the receptive coverage by the American press, Hereward Carrington also saw a moment of opportunity. He invited Eusapia Palladino to tour the eastern United States, hoping that this was the right time for her strange talents—as demonstrated in Naples—to convert American skeptics.

For once, Hyslop and James found themselves in complete agreement. They both thought this an atrocious idea.

Hyslop warned Carrington, who had left the ASPR to pursue his own investigations, to proceed more cautiously. He reminded his former employee of Eusapia's troubled history; he invoked the ease with which Richard Hodgson had exposed her as a cheat. She was untrustworthy, Hyslop said, exactly the wrong kind of medium to parade before skeptics. James also invoked Hodgson's name in recommending against the project. He predicted that the flamboyant Eusapia would attract the worst variety of newspaper attention—exactly the type of publicity that he thought was most harmful to their endeavor.

Further, he thought Carrington was being naive if he thought American scientists wanted an honest investigation of the Italian medium. Long and bitter experience had taught him otherwise. James suspected that researchers would see her merely as an opportunity to further discredit psychical research.

"It would be one thing if the scientists would really investigate Eusapia," James wrote. "But I have very little faith in the candor of such men, and doubt any important result accruing." He hoped Carrington would follow his advice: "Let them [the scientists] perish in their ignorance and conceit."

If this struck an unusually bitter note for James, it was also the sound of recent experience. He and his colleagues had allowed Stanley Hall to have his requested time with Leonora Piper. They were now reaping the unhappy results.

Once he'd achieved his object, Hall made it clear that he had never intended to find out whether Mrs. Piper could communicate with spirits. That concept belonged "more to the troglodyte age than our own." His interest was in debunking her once and for all: "Seriously to investigate the problem of whether discarnate ghosts can suspend any of the laws of matter seems to me not only bad form for any and every scientific man, but an indication of a strange psychic rudiment . . . that ought to be outgrown like the prenatal tail or gills."

What Hall had in mind—in the six sittings he'd scheduled—was a clinical dissection of this vestige of superstitious belief.

Dismissing earlier studies as inadequate, Hall and his research assistant, Amy Tanner, insisted on first retesting Mrs. Piper's trance state. Hall counted her breaths, measured her pulse rate, timed the whole affair. It took about six minutes for her to drop into trance, he reported. At that point her breathing slowed from more than twenty breaths a minute to fewer than ten. Her pulse fell from eighty-four beats a minute to seventy.

It was in this state that Mrs. Piper would begin to scribble messages on paper. She resurfaced slowly, some fifteen to twenty minutes of gradual awakening. As the minutes went on, her breathing quickened, her pulse climbed back to normal.

The trance seemed real enough. But Hall thought he could prove otherwise. He mixed up a spirit of camphor (a slightly toxic compound known for its stinging taste and ability to numb sensation) and dripped it into her mouth. To his surprise, she did not startle awake. But on emerging slowly from the trance, she exclaimed in dismay that her mouth was numb. The next morning blisters covered her swollen lips and tongue. For days she had difficulty swallowing.

Hall's next attempt to dismantle Mrs. Piper's trance involved an esthesiometer, an instrument that allowed him to slowly screw a weight against her skin, testing for the sensation of pressure. He expected that as the pressure increased, it would bring her awake. Again, the experiment failed to disturb the trance. But afterward her hand and arm were pockmarked with red pressure points where the weight had been screwed into her skin. For a time, she feared she'd lost full use of the hand; it tingled and refused to respond for several days.

Alta Piper, now twenty, was furious. Was this what "real science" was all about? Injuring a woman who was, as ever, trying to be helpful? Alta fired off the accusations to Hall, who acknowledged that perhaps "we went farther than we should" and asked to be allowed to continue the sittings. Alta voted no; she wanted the door slammed in Hall's judgmental face. But her mother said yes. She was willing to forgive the first painful experiments, and had "no objections to experiments of any sort if they left no bad aftereffects," Mrs. Piper wrote to Professor Hall.

Leonora Piper was weary of being a mystery, most of all to herself. She clung to the hope that this highly respected psychologist would be able to give her some answers.

MRS. PIPER'S SO-CALLED spirits were a joke, Hall reported.

Her primary control, Rector, fished for information and couldn't tell if it was accurate or not. At one point, Hall invented a dead niece, who then "sent" ghostly messages in later sittings. The so-called Hodgson control was no better, claiming to remember conversations with Hall that had certainly never taken place. Mrs. Piper's trances seemed genuinely odd, yes; occasionally, she revealed information she shouldn't have known; but overall, his diagnosis was that a competent doctor, one experienced in mental health, could have cured her years ago.

"In fine, at the very best, I for one can see nothing more in Mrs. Piper than an interesting case of secondary personality with its own unique features," Hall said. He could only conclude that people found meaning in

her trance utterances because they wanted to find it. He included his colleague William James in that group of self-deceiving dreamers.

Hall hoped that the rest of society would appreciate the short work that
a good scientist could make of even an acclaimed medium. He hoped that
people would see it as he did, as a battle of good science against evil mysticism: "Science is indeed a solid island set in the midst of a stormy, foggy,
and uncharted sea, and all these phenomena are of the sea and not the
land. If there have been eras of enlightenment, it is because these cloud
banks of superstition . . . have lifted for a space or a season.

"Spiritism is the ruck and muck of modern culture, the common
enemy of true science and true religion, and to drain its dismal and mias-
matic marshes is the great work of modern culture."

In the interests of honest reporting, Oliver Lodge wrote to Hall and
asked to publish his study of Mrs. Piper in the SPR journal. The experiments should be kept in context with the others, Lodge pointed out; "the
full moral ought to be extracted from them and they ought to go on record
with the rest. In so far as they have been painful, it is the more desirable
that they not be wasted."

Hall refused. He had other plans for the material, he said, perhaps a
book with Tanner on spiritualism itself; perhaps later publication in "some
journal," perhaps a report or two in his own *Journal of Psychology.* He suspected the SPR, given the chance, would pretty up his findings. As he
wrote to Lodge, Hall wanted to be sure that his own inferences, which "are
pretty negative," would be fully included, and he doubted that would happen in any publication edited by believers.

Lodge sent Hall's letter over to William James with a terse note: "By
this morning's post, I have received from Stanley Hall a letter, which
strikes me as arrogant and unsatisfactory. I enclose a copy and do not wish
a reply."

As expected, Amy Tanner's book *Studies in Spiritism,* published the following year, detailed every fault in the Piper sittings as witnessed by the
author and her mentor. But if Tanner's conclusions were unsympathetic to
psychical research, they were not entirely unsympathetic to Mrs. Piper herself. Tanner's perspectives on the medium and her life as a research subject

were, in their way, kinder than any analysis made by the dedicated psychical researchers, including those who believed that Leonora Piper possessed supernatural abilities.

Even Hodgson had never thought Mrs. Piper intelligent enough to understand her own capabilities, much less the metaphysical questions raised by spirit communication. "Mrs. Piper's opinion, in any case, is of no value," Hodgson once told an ASPR member, explaining why the medium looked to investigators for help with a mystery "which she herself had no hope of solving."

Tanner didn't claim to see an intellectual, but she did see a woman trapped and isolated by the mediumistic side of her life: "She hides from the general public as much as possible," she wrote. As a longtime medium, Mrs. Piper had given up all traces of the friendly little girl from New Hampshire. She was suspicious of strangers. She tried to conceal her work from her neighbors, asking sitters to enter and go upstairs quietly, watching at the door so that visitors didn't even sound the doorbell.

"She was brought up a Methodist, but when her parents moved to a town where there was only a Congregational Church she attended that. She would like to have some church connections," Tanner wrote, "but seems to feel that probably she would not be welcomed in any church on account of her work as a medium. Here too she is isolated."

Mrs. Piper had filled her life instead with her family, with art and music, and with the outdoors, becoming "unusually fond of nature and its beauties." But that hardly balanced the more than twenty years in which she had been singled out, made into a freak, given up hours of her life to trances and tests, demanded by psychical researchers seeking that elusive crack in the wall between the living and dead.

If Tanner and Hall concluded that Leonora Piper had grown a tangle of secondary personalities, who could blame them? Both James and Lodge had already raised the possibility, particularly the concern that the often-domineering Hodgson might have inadvertently induced some of them. They had long suspected that the equally domineering control, Rector—who, unlike Mrs. Piper, could stand up to Hodgson—arose from her subconscious as a direct defense against the investigator.

Amy Tanner also wondered if the Mrs. Piper of 1909 was a worn-down version of a once more talented medium. If so, she thought, the investigators might be partly to blame. But she also pointed out that Mrs. Piper's powers, whatever they were, had been strongest during earlier times of poor health.

During the period when the medium suffered severe abdominal pain, eventually diagnosed as an ovarian tumor, her readings had been extraordinarily impressive. G.P. had appeared during the worst of that illness, including the time when Mrs. Piper was recovering from the surgery. He had faded away only as her health returned. Tanner speculated that perhaps in a physically fragile state Mrs. Piper became more of a mental conduit: "That is, the facts in the case seem to point to the theory that the mediumistic power is encouraged and, perhaps in the beginning, caused by nervous shock."

In her way, it seemed, Dr. Amy Tanner was raising the possibility of the decline effect, the recognition that no psychic, even the best one, lasted forever.

SAILING ACROSS THE Atlantic to New York, Eusapia Palladino decided to hold a séance to alleviate her boredom. Passengers gathered round, clamoring to talk with their dead relatives. A woman fainted to the floor as lights winked in the air.

Onboard journalists cabled the story of the carnival-like spectacle to New York, where Hereward Carrington immediately received offers from music hall managers to host Eusapia. Deep in debt on the venture, Carrington quickly accepted the most lucrative among the invitations.

"Poor Carrington had to promise her enormous pay, and to raise the money he had to give sittings to every idle rich person who asked for them, hoping to invite some serious experts gratis with the surplus," James wrote to Flournoy with some sympathy.

Hyslop felt no sympathy at all, only alarm. Everybody knew that Eusapia had to be tightly managed, or she resorted to cheating. As far as Hyslop could tell, this tour promised none of the control and all of the opportunity for her to make a complete fool of herself, and of psychical researchers as well.

Although he had refused money to Carrington earlier, Hyslop now anted up for a stenographer in an effort to at least make records of the sittings. He also did his best to lower expectations for Eusapia's visit, warning newspaper and magazine reporters that this was an exhibition rather than a scientific project. Hyslop wished she'd stayed in Italy. He wished he was still giving interviews about the cross-correspondence studies. He had an idea that their small set of credible findings was about to disappear in the uproar of an ectoplasmic circus.

A former colleague of James, Harvard psychologist Hugo Munsterberg, was among the first scientists to wangle an invitation to observe Eusapia. He was chosen to attend one of those select séances that Eusapia was giving to wealthy donors.

Munsterberg and James had once been good friends, and James had helped arrange the German scientist's move to Harvard in the late 1890s. He'd also endorsed Munsterberg as president of the American Psychological Association in 1898.

But their friendship had eroded over differences in personality, philosophy, and attitudes toward psychical research. Munsterberg possessed an orderly and applied view of psychology. The philosophy of it didn't appeal to him as much as making it a useful science. He researched eyewitness testimony with an idea toward improving criminal investigations; he studied the ways in which doctors could use psychology to help patients, such as encouraging them to believe that they were getting well.

As James moved deeper into metaphysical questions of philosophy, Munsterberg responded with disdain and accusations that his colleague was turning away from real science. He'd disliked the books James published since retirement, *Pragmatism* in 1907 and *A Pluralistic Universe* in 1909. Both of them discussed philosophy's potential to explore truth and reality in a way that science—even good applied science—could not.

The philosophy of pragmatism had gained a following among young philosophers who saw it as way of shedding nineteenth-century romanticism for twentieth-century realism. Among other things, James's philosophy raised the idea that real-life experience might be as important to interpreting life as scientific absolutes: "On pragmatic principles," James explained, "if the hypothesis of God works satisfactorily in the widest sense of the word, it is true."

Munsterberg made a point of mocking James's new philosophy when he spoke publicly. He was so antagonistic that James's wife, Alice, began taking the attacks personally. Alice would barely speak to Munsterberg—she was "pure ice," as her husband described it—when she met their former friend at Cambridge social events. James himself wrote to Munsterberg in dismay: "Were it not for my fixed belief that the world is wide enough to sustain and nourish without harm many different types of thinking, I believe that the wide difference between your whole *Drang* in philosophizing and mine would give me a despairing feeling. . . . I am satisfied with a free wild Nature; you seem to me to cherish and pursue an Italian Garden, where all things are kept in separate compartments, and one must follow straight-ruled walks."

Like so many in the American psychology community, Munsterberg deplored James's detour into psychical research, and felt compelled to respond and protect the profession from the taint of such pseudoscience. When Carrington issued him an invitation to one of the first sittings with the infamous Italian medium, Munsterberg accepted with pleasure—and an agenda.

He seated himself next to Eusapia at the December 18, 1909, séance, holding her left hand and with her left foot resting on top of his polished shoes. He did not try to hold her foot down; she quickly slipped out of the shoe and wiggled her foot backward to move a small table behind her. Another guest grabbed her toes; she began to screech and halted the séance. Munsterberg promptly contacted the New York papers, claiming that he arranged for the medium to be caught cheating. When the other sittings with American-based scientists proved equally disastrous, Munsterberg told the *New York Times* that the gullibility and susceptibility of the previous investigators explained, entirely, her reputed ability to do magic.

"There is no limit to his genius for self-advertisement and superficiality," James fumed to Flournoy. "Mendacity too!"

In a follow-up article for the *Metropolitan Magazine,* titled "My Friends, the Spiritualists," Munsterberg lamented the delusions of otherwise intelligent men, naming Ochorowicz and Richet as spiritualists in the most credulous sense, and, as James wrote, in another angry letter to Flournoy, attempting to "insinuate that I also am one."

The irritating thing, Flournoy replied, was that all of these cheats had been demonstrated before; anyone who knew Eusapia knew that she liked to cheat when she could. "Fraud with the feet, which Munsterberg allowed to take place (if indeed he didn't intentionally encourage it in order to support his preconceived ideas) does not explain the innumerable other happenings which occur when she is controlled in all 4 limbs, and the séances take place in sufficient light."

Agreed. And yet James feared that Hyslop's warnings had been proved correct, that nothing good would come of this particular expedition. "Eusapia's trip to the U.S. will simply have spoiled her, and discredited everyone else," he predicted. His own reputation had suffered already. An editorial in the *Times* made that more than clear in its concluding sentences: "We are not so much concerned about Eusapia Palladino as we are about the psychical researchers. . . . The time has almost come when to be a psychical researcher is to confess unsoundness of judgment."

The editors seemed to question the judgment of two people in particular, both of whom were named in the editorial: Hereward Carrington—who should have known better than to bring over this ludicrous fraud—and William James, who should have known better throughout his career.

THAT SAME STORM-CLOUDED FALL, an essay by William James appeared in the *American Magazine*. Called "The Confidences of a Psychical Researcher," it had been intended as a retrospective but served, serendipitously, as an eloquent answer to his critics.

James began by recalling Henry Sidgwick, with his shy stutter, his "liberal heart," his rare combination of "ardor and critical judgment," and his complete frustration over the elusive nature of psychical phenomena. "I heard him say, the year before his death, that if anyone had told him at the outset that after twenty years he would be in the same identical state of doubt and balance that he started with, he would have deemed the prophecy incredible.

"My own experience has been similar to Sidgwick's."

After twenty-five years working with some outstandingly good psychical researchers, conducting experiments, studying the literature, sitting with mediums both gifted and fraudulent, James found himself stymied. He could accept some of the phenomena as real, but he could not explain them.

"I confess that at times I have been tempted to believe that the Creator has eternally intended this department of nature to remain *baffling*, to prompt our curiosities and hopes and suspicions, all in equal measure, so that, although ghosts and clairvoyances, and raps and messages from spirits, are always seeming to exist and can never be fully explained away, they also can never be susceptible of full corroboration."

James deplored the apparently incurable dishonesty associated with spiritual endeavors and the way that it continually obstructed progress. Psychical researchers routinely wasted valuable time exposing cheaters rather than studying legitimate phenomena. From Richard Hodgson's dissection of Madame Blavatsky to Everard Feilding's hilarious encounters with spirits of the still-living, the potential for fraud appeared infinite.

Frank Podmore, who sometimes sardonically referred to himself as the SPR's "skeptic in chief," had published a two-volume history of spiritualism, suggesting that against such a background all mediums *must* be considered suspect, dismissing Eusapia Palladino as a bad joke played on his colleagues and Leonora Piper as a woman with some telepathic skills and an excellent memory for facts shared casually by her sitters. He had no proof of the latter, Podmore said, but her overall record, although impressive, failed to convince.

Perhaps this was too cynical, Podmore allowed. "The accurate appreciation of evidence of this kind is an almost impossible task," he wrote in his book *Modern Spiritualism.* "Mrs. Piper would be a much more convincing apparition if she could have come to us out of the blue, instead of trailing behind her a nebulous ancestry of magnetic somnambules, witchridden children, and ecstatic nuns."

The same ancestry made Eusapia Palladino impossible to defend, Podmore said flatly, although Hereward Carrington was willing to try. Drawing on his own talents as a magician, Carrington had tried to repair her reputation—and his—by holding a New York stage show to illustrate the difference between conjuring—which he knew well—and real magic.

As reported in city newspapers, he began by stepping onto the stage of New York's Berkeley Theatre and placing a wax hand atop four glass tumblers sitting on a wooden table. He then retired to a corner of the stage and "asked" the hand questions. In response, the wax fingers had rapped on the glasses. The audience was mystified, reporters declared, until Carrington revealed a fine black thread attached to the hand, and an assistant attached to the thread, hidden behind a curtain, busy tugging on the line.

Carrington levitated tables using wires and hooks that were concealed in his sleeves, materialized a floating baby's hand (attached to his foot), generated from a cabinet a misty figure which turned out to be a piece of cheesecloth dusted with phosphorus. He knew the tricks, Carrington said, and he'd checked for all of them with Eusapia Palladino. "I have always said that she will resort to trickery if she can, but if she was carefully watched she still performs the most marvelous acts and some of these acts I can explain only on supernormal grounds." He saw, belatedly, that he had underestimated his opposition, that unsympathetic researchers had deliberately ignored his warnings, wishing only to see the medium exposed.

At Carrington's request, Everard Feilding returned to Italy to retest Eusapia after the American tour. He found her sick, demoralized, viciously bitter, and unable to produce a single decent result: "Everything this time was different," he said. If she had ever possessed an unpredictable power, it had abandoned her.

Neither Hyslop nor James rejoiced that their warnings against her American tour had proved true. They would much rather have been wrong.

PERHAPS, James wrote in his last essay on psychical research, it was unfair to expect anything resembling purity in the endeavor. All human enterprises contained some fraud; James accepted that it was an integral part of human nature to sometimes prevaricate, to wander that fine line between true and false, right and wrong.

"Man's character is too sophistically mixed for the alternative of 'honest or dishonest' to be a sharp one," he said, noting that despite its attitude of superiority, science itself was not immune to deceit. "Scientific men themselves will cheat—at public lectures—rather than let experiments obey their well-known tendency toward failure." He recalled a well-known physics demonstration using an apparatus intended to show that whatever the outer stresses, its center of gravity remained immovable. When a colleague borrowed the device, though, he found it wobbled through his demonstrations. Well, explained its owner, "to tell truth, whenever I used that machine, I found it advisable to drive a nail through the center of gravity."

Secretly stabilizing the device did not undo the laws of gravity, James pointed out; in fact it helped audiences understand the scientific point rather than subverting it. By the same token, fraud among professional mediums did not undo the possibility of real supernatural phenomena. Perhaps, curiously, fraud might serve to corroborate truth. "If we look at human imposture as a historic phenomenon, we find it imitative"; tricksters were only able to garner attention because they faked something that did exist, fraudulent mediums only persuasive by taking advantage of the reputation earned by the few legitimate psychics. "Those who have the fullest acquaintance with the phenomena admit that in good mediums *there is a residuum of knowledge displayed* that can only be called supernormal; the medium taps some source of information not open to ordinary people."

That also seemed to be Podmore's opinion, that "even the extravagances of mysticism may contain a residuum of unacknowledged and serviceable fact." It was an idea culled from Gurney and Myers's original calculation that perhaps 5 percent of all the occult claims they studied were real, and all these years later, their old colleague continued to push against "scientific rejection" of the legitimate phenomena. So did William James make a stand on that narrow ground.

"Either I or the scientist is of course a fool," James wrote, "with our opposite views of probability here. . . . I may be dooming myself to the pit in the eyes of better-judging posterity; I may be raising myself to honor; I am willing to take the risk, for what I shall write is my truth, as I now see it."

First, he believed that these odd, these occult, these so-called supernatural phenomena occurred with remarkable frequency, despite scientific claims that they were "so rare as to be unworthy of attention." Second, James believed in "the presence, in the midst of all the humbug, *of really supernormal knowledge,* beyond the ordinary senses."

And finally, regretfully, James believed he and his colleagues had been "too precipitate in their hopes," had trusted too much in the ability of science to solve all mysteries. The answers would not come in his lifetime, he suspected, and perhaps not in his children's lifetimes either.

MARK TWAIN WAS born in 1835, a year in which Halley's Comet had blazed like God's own lightning across the night skies. Now, in 1910, the comet was scheduled to make its return, even closer to Earth, and, so the stories went, dragging death in its glowing wake.

Pharmacists began selling "comet pills" to protect against poisonous gases that might accompany Halley's Comet. In New York and other big cities, stores sold out of telescopes needed to track the fire in the sky. Churches held prayer services, people gathered on rooftops to watch for the comet's approach.

And Twain, now seventy-five years old, was thinking not only of the cycle of the comet, but of the cycle of his life as well: "I came in with Halley's comet. . . . It is coming again . . . and I expect to go out with it . . . the Almighty has said, no doubt: 'Now here are these two unaccountable freaks; they came in together, they must go out together.'"

Twain died on April 21, 1910, one day after Halley's comet reached its closest point to Earth, while "comet parties" danced on rooftops around the world, while reformers preached that the comet was a warning, a judgment on the godless twentieth century. When King Edward VII died on May 6, following a series of heart attacks, newspapers noted that the comet's path was exceptionally erratic that day.

The comet's reputation as a harbinger of death shone so brightly that Pope Pius X felt required to reassure the world's faithful. As the pope would remind his followers, the comet's tail had missed Earth by almost 200,000 miles. He trusted in the judgment of astronomers when they called Halley's Comet an interesting celestial phenomenon, not evidence of God's disappointment with the modern world.

ON AUGUST 19, 1910, Frank Podmore, the last of the three authors of *Phantasms of the Living*, was found dead, floating in a small pond in the resort community of Malvern Wells.

Maintaining his role as in-house skeptic for the SPR, Podmore had recently infuriated his colleagues by suggesting that Eusapia Palladino's dazzling performance in Naples, so different from the American tour, undoubtedly occurred with the help of an accomplice. After sending Carrington, Baggally, and Feilding into a frenzy of outrage and denial, Podmore had left for a golfing holiday. He'd spent a relaxing week at Malvern Wells and on a mild Sunday night gone out for a late-night walk, stopped for a cheerful conversation with a friend, and vanished. His body was found five days later in a small pond; at a same-day inquest, the coroner gave a simple verdict of "found drowned," calling the death an unsolved mystery.

"Suicide has, of course, been suggested," John Piddington wrote to James Hyslop, "but there is no proof of it and I see no hope of the mystery ever being cleared up."

The previous year, Podmore had separated from his wife and resigned his longtime job as postal inspector amid a flurry of whispers that he'd been caught in a homosexual relationship. Piddington warned Hyslop to watch out for any references to such "grievous trouble" in séances yet to come, and begged him not to publish anything sexually revealing that "may be said about him or purport to come from him in script or trance.

"At the same time, I must not let you get the impression . . . that there is warrant for connecting his death with his troubles. There may have been a connection but there is absolutely no evidence of any. I know I can rely upon you to regard this letter as absolutely private and confidential and I think you would do well to destroy it after reading it."

Hyslop kept the confidence. But he kept the letter too.

NO ONE FROM the SPR's office came to the small private burial for Podmore; as his friends noted resentfully, the association didn't even send a wreath. But Nora Sidgwick wrote a heartfelt appreciation in the society's *Proceedings:* "Ignorant criticism we can get plenty of, but when not harmful it is usually quite useless. What is not easy is to find a man with unflagging energy in keeping his knowledge up to date, unflagging belief in the importance of the investigation, who yet can put himself outside it.

"The Society will be fortunate indeed if it finds another critic equally friendly, learned, painstaking and accurate . . . to put the brake on where there are signs of running too fast." As she reminded the membership, skeptics were as important to making a convincing case as optimists were. Perhaps more so.

WILLIAM AND ALICE JAMES left England near the time of Podmore's death. They had arrived in the spring to stay with his brother, Henry, who had fallen prey to illness and depression and wanted company.

The Jameses had barely settled into Henry's comfortable home in Rye, where he had moved from London, when a telegram arrived bearing bad news. Their youngest brother, Robertson James, was dead of a heart attack in his sleep. William's reaction was partly grief, partly envy. His heart disease had worsened; he could hardly bear to take a step, even to breathe. He wished he would go out so easily, he said.

By August, William was so sick that he begged to go home. Alice and Henry, who had returned to good health, booked a voyage back to the United States and, once there, headed directly to Chocorua. William immediately took to bed, and the doctors they consulted predicted that he would gradually recover.

Within the week, though, William was so weak he couldn't eat, could barely keep down a few swallows of milk. Early on the afternoon of August 26, Alice came into the sickroom and found her husband unconscious. She climbed into the bed and held him against her, listening to his painful breathing until there was no sound at all.

An autopsy, requested by Alice James, showed that the cause of death was acute enlargement of the heart. But even so, she believed that his will had played a role. "He wanted to go," Alice said, "and departed swiftly as he always has when he made up his mind to move on."

In the United States and Europe, newspapers announced James's death with ceremony, reporting the loss of "the most distinguished and influential American philosopher of our day." His family and friends simply grieved for the man, "the whole unspeakably vivid and beautiful presence of him," as his brother Henry wrote in memoriam.

In the months after William's death, Alice and Henry met with several mediums, she, particularly, hoping for a message that he lived on. They did not have a sitting with Leonora Piper, who had declared a retirement after her painful encounter with Stanley Hall. Their sessions with other Boston mediums conveyed nothing, Henry said, but the grim refusal of the dead.

Alice was disappointed, but her faith was unshaken. She had always liked William's idea that "a will to believe" was the most important part of living in a spiritual universe. "I believe in immortality," she wrote to a friend. And she also believed that William was "safe and living, loving and working, never to be wholly gone from us."

The spiritualist community, however, wanted more, or at least wanted more of a demonstration. Within days, newspapers carried multiple claims of contacts with the spirit of William James.

A Boston businessman told the *New York Times* that James had "sent a message to his friends from the spirit world" during a séance with a most respectable medium. The message was rather vague: "I am at peace, peace—with myself and all mankind. I have awakened to a life far beyond my highest conception while a denizen of earth." But the "spirit" promised to contact Henry James shortly with more explicit details.

The pastor of Boston's Unity Church announced that he had felt "spiritualist vibrations" which he thought came from James because they were intense and rapid, indicating "a genius wishing to communicate rather than a common soul."

James Hyslop reported that one medium had relayed some messages from James, in one case concerning a private conversation between Hyslop and the late psychologist, which was correct in all details, right down to the garments they'd been wearing. But this was suggestive only, Hyslop added; he hadn't seen anything in the way of satisfactory proof. He hoped that the fascination with whether William James could come back from the dead wouldn't obscure the more important facts at hand: that psychical research had lost one of its best friends, that there was much work yet to be done.

The *Times* decided to ask an expert to settle the matter, the famed scientist-inventor, Thomas Alva Edison, currently working to turn silent motion pictures into talking ones.

Edison's name was almost synonymous with the power of modern science and technology, and as the newspaper proudly revealed, this was his first and only interview on the subject of survival after death. "The occasion was the recent death of Prof. William James, Harvard's distinguished psychologist, and the alleged reappearance of Prof. James's soul on earth. The newspapers have been teeming with the subject. The psychic researchers are even now quarreling bitterly over it. The public is puzzled.

"Therefore I turned to Edison," the reporter explained, "who has solved for us so many puzzling problems."

Edison saw no particular puzzle here. He didn't believe in immortality because he didn't believe in the human soul, he told the *Times* briskly. Or as the headline said: "Human Beings Only an Aggregate of Cells and the Brain Only a Wonderful Machine, Says Wizard of Electricity."

It was the mechanical universe that formed the basis of Edison's belief—planets spinning, winds blowing, people born and dying—all simply carried onward by that well-oiled machinery of creation. There was no reason to believe that the human brain would continue after death, Edison pointed out, any more than to think that one of his phonographic cylinders would be immortal. No machine—no cog in the works, as humans appeared to be—would live forever.

"No, all this talk of an existence for us, as individuals, beyond the grave is wrong. It is born of our tenacity of life—our desire to go on living—our dread of coming to an end as individuals. I do not dread it though. Personally I cannot see any use of a future life."

But then, Edison thought the newspaper was asking a dated question, relating to a past time when people believed in a personal God, burned their candles, and kept their faith for an Almighty who spoke to them of morality and decency and a better life beyond this one. Edison saw no evidence of such a supreme being, or such moral behaviors, in this life any more than the next one.

"Mercy? Kindness? Love? I don't see 'em. Nature is what we know. We do not know the gods of the religions. And nature is not kind, or merciful, or loving. If God made me—the fabled God of the three qualities of which I spoke: mercy, kindness, love—He also made fish I catch and eat. And where do His mercy, kindness, and love for that fish come in?"

If a determined reader continued to the end of that long interview, there was a hint that Edison didn't see the world as quite such a finished machine as he had indicated at first. Running like a small rough thread of doubt through his polished confidence was the faintest glint of humility.

"Now I am going to ask you a question," Edison said to the reporter. "Why are you here for—here on earth, I mean?"

As the journalist confessed in the newspaper article, he had no good answer to that question.

"Well, there you are. We do not understand. We cannot understand. We are too finite to understand. The really big things we cannot grasp as yet."

It was exactly the kind of point that William James would have agreed with, the kind of discussion he would have enjoyed. But the newspaper put

Edison's admissions of uncertainty at the end of the story. They didn't fit neatly into an account of the ways the twentieth century had left dusty notions of faith and spirit behind.

IT HAD BEEN barely a year since Oliver Lodge had announced, also in the *Times,* that he and his colleagues had found evidence to link the dead and the living. In the forgotten hopefulness of Lodge's predictions, in the clear confidence of Edison's commentary, one could read between the newspaper lines and see, far better than in any light-shot crystal ball, the look of the future.

Oliver Lodge would continue to argue the case for life after death, through the coming decade and beyond. His argument would turn more personal after one of his sons, Raymond, died in battle during World War I. Nora Sidgwick would maintain her insistence on objectivity, but concede in 1913 that the cross-correspondence studies offered real evidence of "cooperation by friends and fellow-workers no longer in the body." Charles Richet would continue dividing his time between traditional physiology and unorthodox investigations of the occult, insisting that the best hope in resolving difficult questions was for good scientists to tackle them: "Our duty is plain. Let us be sober in speculation; let us study and analyze facts; let us be as bold in hypothesis as we are rigorous in experimentation. Metaphysics will then emerge from Occultism, as Chemistry emerged from Alchemy; and none can foresee its amazing career."

Carrington, Baggally, and Feilding would continue their work as psychic investigators for many years—debunking a number of well-known psychics along the way. Of the three, Carrington would achieve international fame as a psychic detective; the American magician Harry Houdini once called Carrington's writings on the subject the best ever produced. Carrington and his colleagues, though, would always regret the collapse of their work with Fusapia Palladino, who died in 1918 due to complications of diabetes. They all agreed, though, with Feilding's assessment, in a letter to Carrington, that her reputation could never have been restored: "The public is what it is, scientific people are what they are, and nothing can be done."

James Hervey Hyslop would lead the American Society for Psychical Research—almost single-handedly keeping it in the public view by force of personality—until his death in 1920. His passion for evidence and argument would remain unabated. As he wrote in the year of Palladino's death, "Any man who does not accept the existence of discarnate spirits and the proof of it is either ignorant or a moral coward. I give him short shrift, and do not propose any longer to argue with him on the supposition that he knows anything about the subject." His longtime secretary would later write her own book, *James H. Hyslop X: His Book,* describing the return of his spirit, distinctive in its crisp personality and outspoken demeanor.

Leonora Piper would return to work as a medium (and outlive almost all her investigators, dying in 1950 at the age of ninety-three). She would not, however, outlive the debate over matters of life and death, science and religion, a debate that the pronouncements of even the great Thomas Alva Edison could not begin to resolve.

Among Christians there would be a growing, militant opposition to a reality defined only by secular values—growing in strength after World War I, as many, especially Americans outside of the major eastern cities, longed for a return to prewar innocence. (Warren G. Harding, promising a "return to normalcy," won the U.S. presidency in 1920.) Fundamentalist creationists would begin their battle against the teaching of Darwinian evolution in the U.S. public schools, a battle highlighted by sensationalistic media coverage of high school biology teacher John T. Scopes's 1925 trial in Dayton, Tennessee. The charge: violating a Tennessee state law banning any teaching that contradicted the divine creation of man as described in the Bible.

Far from disappearing in favor of scientific materialism, spiritual values would endure and even seem to gain ground in the aftermath of the Great War, even in intellectual circles—as evidenced, for example, by the devout Christianity of twentieth-century authors J. R. R. Tolkien and his friend C. S. Lewis, both of them battle veterans and Oxford dons.

In 1926, Everard Feilding would review the very human needs and experiences that kept spiritual beliefs—or perhaps hopes—alive, criticizing both the establishments of science and religion for failing to recognize their importance.

Most people, Feilding would write, were "unwillingly children of the time in which they live." They lived surrounded by new knowledge, inundated by facts; they were told absolutely that such information was the only route to certainty about the universe. They were given no guidance as to how religious feeling, faith, or intuition might fit into that world; they were given less guidance if they experienced a supernatural event—saw a crisis apparition, had a premonition, or simply felt an inner sense of belief in something more. "If but some link could be established between the two, some stepping-stone laid on which they could venture out into the dark stream, their confidence would be restored," Feilding would insist. And he would mourn the past, grieve for the loss of that moment when he and his friends had thought they might reconcile science and faith after all, and find that elusive path, as faint and as real as moonlight, leading to a universe in which all things were possible.

ACKNOWLEDGMENTS

I'VE NEVER HAD premonitions, eerie dreams, heard voices, saw ghosts, or possessed any sense beyond the basic five, and I've never wished to acquire such talents. I was always the kid who pushed the Ouija board pointer out of sheer boredom, and to this day, I believe that it would never have moved without a helpful shove.

So when I started this book, I saw myself as the perfect author to explore the supernatural, a career science writer anchored in place with the sturdy shoes of common sense. In the way that books do—or that one hopes they do—this one changed the way I thought, and definitely altered that sense of perfection. I still don't aspire to a sixth sense, I like being a science writer, still grounded in reality. I'm just less smug than I was when I started, less positive of my rightness.

What changed? I had the pleasure and privilege of spending three years in the company of genuinely brilliant thinkers—William James and his colleagues who questioned and explored possibilities so acutely that it was impossible not to reevaluate my assumptions. I participated in a slightly unnerving ESP experiment. I read reports by psychical researchers that I couldn't explain away. I thought all over again about the shape of the world, about science, about the limits of reality and who sets them, illuminated by history, philosophy, theology as well as science. There were days when I could feel the hinges of my brain, almost literally, creaking apart to make room for new ideas.

Before I get into specific thanks, I want to first express my gratitude to the many people who, in the most casual and everyday sense, also forced me to look beyond the accustomed horizons of my life. I would like to thank the secretary in my department at the School of Journalism and Mass Communication at the University of Wisconsin-Madison, who trusted me with the story of her haunted house. The scientist from Stanford who believed in and practiced ESP, the prenatal care nurse who regularly consulted a medium, the physicist from a Florida university, who fled a ghost-ridden laboratory, the music store clerk who saw a specter, the countless people who told me their stories, once they heard the subject of my current book. Ordinarily, science writers don't get told personal ghost stories. I suspect that's because we are regarded, rightly, as an unsympathetic audience. But once I took on this particular subject, I realized—as the people I write about also learned—how strong and true this current of belief and experience runs in our society. I won't tell you that I suddenly believed in every ominous shadow that loomed in these encounters. But they formed a very curious pattern. I would specifically like to thank my terrific father-in-law, David Haugen, who experienced—and recently told me—the best "crisis apparition" story I ever heard.

I would also like to thank my editor, Ann Godoff, who believed in this story from the beginning, propped me up through all the worst parts of the telling, and made the work better at every turn. The people at Penguin Press are amazing to work with and I would especially like to thank Liza Darnton and Beena Kamlani, who made me look smarter than I am, and Sarah Hutson. As ever, I am grateful to Suzanne Gluck, who combines being the best agent I know with being one of the smartest and nicest women in my life.

I would like to thank the patient and welcoming staff of the American Society for Psychical Research, who set me up at a library table in their beautiful brownstone and allowed me to read through letters from the early history of their organization. I would especially like to thank ASPR executive director, Patrice Keane; archivist, Colleen Phelan; librarian,

Jeremy Shawl; and former librarian, Grady Hendrix, for their kindness and meticulous research help in this regard. I appreciate the permission to quote from those letters, also the kindly permissions from the British Society for Psychical Research, and its helpful librarian, Willis Poynton; the Houghton Library of Harvard University, with special thanks to Leslie Morris and Peter Accardo at Harvard, the Bodleian Library of Oxford University, and particular thanks to librarian Colin Harris and the Trinity Library of Cambridge University.

I had a hardworking group of University of Wisconsin students who helped with the research in various stages and I would particularly like to thank Ben Sayre and Rena Archwamety for their help in tracking down nineteenth-century newspapers and magazines; Christine Lagorio, for her study of psychologist Joseph Jastrow; Edna Francisco, Amanda Novak, and Margaret Menge for their investigation of the occult in Victorian times. The University of Wisconsin-Madison provided me both with a sabbatical and summer salary to work on the book and has proved ever supportive of my work as a writer.

It is my pleasure to thank the writers in my life, especially my husband, Peter Haugen, a busy and talented popular historian, who took the time out of his own work to read and improve this story and to argue with me about metaphysics, refusing to ever let me settle for the easy answer. I would also like to express my gratitude to Kim Fowler and Robin Marantz Henig, who were unstintingly generous in both time and helpful suggestions regarding the chaotic early versions of my manuscript.

Here's to my best book friends also: Denise Allen, Pam Ruegg-Morgen, Sue Brown, Julie Hunter, Linde Patterson, Mirriam Rosen, Susan Isensee, Suzanne Wolf, Jody Haun, Jacquie Hitchon McSweeney and Jean Carlson, who cheered me right to the finish.

And as always, here's to Marcus and Lucas, who never let me forget what matters most.

Deborah Blum
Madison, Wisconsin
February 23, 2006

NOTES AND SOURCES

I WAS FORTUNATE to spend time in two terrific and very different archives: the Houghton Library, at Harvard University, which holds the correspondence of William James (referred to hereafter as Houghton), and the American Society for Psychical Research, in New York (referred to as ASPR), which holds a treasure trove of largely unpublished correspondence and other documents relating to James, Richard Hodgson, James Hyslop, and colleagues, as well as containing one of the best occult libraries in the world. Many of the described interactions in this book are drawn from the archived correspondence in those institutions.

As a point of reference, letters from and to William James have, of course, also been excerpted and published many times over; the best companion to the original letters at the Houghton Library is an annotated series of twelve volumes, *The Correspondence of William James* (Charlottesville: University Press of Virginia, 1992–2004). Some of James's more noteworthy correspondence on psychical research and the vast majority of his published articles on the subject are contained in two books: Gardner Murphy and Robert O. Ballou, eds., *William James and Psychical Research* (New York: Viking Press, 1960); and Frederick H. Burkhardt and Fredson Bowers, eds., *Essays in Psychical Research* (Cambridge, Mass.: Harvard University Press, 1986).

For a general overview of the period and the players, I found the following books most helpful: Frank Podmore, *Mediums of the Nineteenth Century* (Hyde Park, N.Y.: University Books, 1963; originally published in 1902 as *Modern Spiritualism*); Brian Inglis, *Natural and Supernatural: A History of the Paranormal* (London: Hodder & Stoughton, 1977); Janet Oppenheim's fascinating book *The Other World* (Cambridge, England: Cambridge University Press, 1985); Alan Gauld's wonderfully down-to-earth history *The Founders of Psychical Research* (London: Routledge & Kegan Paul, 1968), and Sir Arthur Conan Doyle's highly biased and terrifically gossipy and readable *History of Spiritualism,* published in 1926 by George H. Doran, New York. I also found the *Encyclopedia of Occultism and Parapsychology,* edited by Leslie Shepard (New York: Gale Research/Avon Books, 1978), to be a great paranormal trivia resource.

In cases where information is widely known and found in numerous sources, I have not provided specific references. I have, however, occasionally attempted to give a sense of the range of references used in portraying a particular psychic or psychical researcher. And I have occasionally tried to give additional context to a particular moment in history or to explain a reference itself. I have not provided citations for every brief quote, but only for the longer ones. And as a further point of clarification, I occasionally provide narratives in the book, mostly ghost stories and accounts of sittings with mediums. Although those are derived from documents of the time, to be referenced below, their telling here is my own.

Prelude

The first narrative: This story comes from William James's report "A Case of Clairvoyance," *Proceedings of the American Society for Psychical Research* 1 (1907): 221–36. As James noted in the introduction, the research for this report was done in 1898 (interviews were largely gathered by a cousin of James's wife, Harris Kennedy), and the report should have been published the following year. But due to the financial problems then ongoing in the ASPR, the organization did not resume publishing its journal until 1907, when James Hyslop headed the group. This report was published in the first volume of the renewed journal.

5: "had been a brilliant scientist": The British pragmatist philosopher Ferdinand Canning Scott Schiller, a friend of James, told this anecdote about scientific hostility toward psychical research; it is recounted in "Some Logical Aspects of Psychical Research," in *The Case For and Against Psychical Belief,* ed. Carl Murchison (Worcester, Mass.: Clark University, 1927), 215–28. This book was the result of a conference on the issue; participants also included Sir Arthur Conan Doyle, Joseph Jastrow, and the magician Harry Houdini.

1. The Night Side

Biographical information on William James—his character and that of his family, his childhood and the general nature of his upbringing—is widely available. This chapter primarily drew from three books. The first is Howard M. Feinstein, *Becoming William James* (Ithaca, N.Y.: Cornell University Press, 1984). I relied on it in particular for information on the childhood of Henry James Sr. and for its trenchant analysis of the terrible accident that took his leg (pp. 39–43) and its influence on his further life. My favorite traditional biography is Linda Simon, *Genuine Reality: A Life of William James* (New York: Harcourt Brace, 1998), and my favorite intellectual biography is Louis Menand, *The Metaphysical Club: A Story of Ideas in America* (New York: Farrar, Straus and Giroux, 2001). Ralph Barton Perry, *In the Spirit of William James* (New Haven, Conn.: Yale University Press, 1938), does a wonderful job of comparing William James to other philosophers of his time.

7: "like being in the dentist's chair": WJ to G. Stanley Hall, Houghton.

8: "priggish, sectarian view of science": WJ to James McKeen Cattell, printed in *Science,* May 4, 1898.

11: Swedenborg's life is explored to varying degree in the general books on spiritualism listed in the introduction to this section. In addition, there is an excellent short biography of the Swedish mystic in Eric Dingwall, *Some Human Oddities: Studies in the Queer, the Uncanny, and the Fanatical* (Hyde Park, N.Y.: University Books, 1962), 11–68, which was my primary source for the description of Swedenborg's fire vision as investigated by Immanuel Kant.

12: "vermin revealing themselves": Podmore, *Mediums of the Nineteenth Century,* 241.

14: The Night Side of Nature: Catherine Crowe and the influence of *The Night Side of Nature* are referenced in every book on the early history of spiritualism. In his book *Ghosts, Demons and Henry James: The Turn of the Screw at the Turn*

of the Century (Columbia: University of Missouri Press, 1989), literary scholar
Peter G. Beidler calls this work "the most influential single book about ghosts
in the second half of the nineteenth century." A 2000 reprint of Crowe's book
by Wordsworth Editions, Ware, England, contains a thoughtful introduction
by Gillian Bennett, editor of *Folklore.*

16: a couple of farm girls: The life story of the Fox sisters has been dissected
almost since they were born, and every book on the history of the supernatural
discusses them. R. Laurence Moore, *In Search of White Crows: Spiritualism,
Parapsychology, and American Culture* (New York: Oxford University Press,
1977), on the history of spiritualism in the United States, contains an excellent
description of the reaction of the clergy to the Fox sisters. For very different
viewpoints, I also relied on Frank Podmore's cynical look at the Fox family in
Mediums of the Nineteenth Century and Arthur Conan Doyle's detailed account
of their career in *The History of Spiritualism,* 60–118, which includes their
adoption by P. T. Barnum and a description of the sitting with James Fenimore
Cooper.

20: table talking: Entire books are devoted to table talking and its place in Victorian spiritualism; see, for instance, Ronald Pearsall, *The Table Rappers* (New
York: St. Martin's Press, 1972).

20: A letter in the *Times* of London: Michael Faraday's letter to the *Times* was
printed on June 30, 1853. He expanded that letter into the article "Professor
Faraday on Table-Moving," *Atheneneum,* July 2, 1853, 801–3.

22: the impossible, unearthly Daniel Dunglas Home: Home is one of the best-
known (and written about) mediums of the nineteenth century. Here and else-
where in the book where I discuss his story, I relied particularly on Podmore,
Mediums of the Nineteenth Century, 2:205–44; Inglis, *Natural and Supernat-
ural,* 225–39; and William Barrett, *On the Threshold of the Unseen* (New York:
E. F. Dutton, 1917), which describes Home as "the most remarkable psychic
ever investigated" (57). The descriptions of the religious overtones with which
Home infused his sessions and the general quality of his séances come from
Conan Doyle, *History of Spiritualism,* 1:187–207; and Hereward Carrington,
A Primer of Psychical Research (London: Rider, 1932). The attitude of Robert
Browning toward D. D. Home is thoroughly discussed in Dingwall, *Some
Human Oddities,* 191–228, with a strong implication that Browning was as
disgusted by rumors of Home's homosexuality as by any possibility that he
cheated as a medium.

23: Ira and William Davenport: The Davenport Brothers are a staple of books on early spiritualist history. I particularly liked the snide description of the Harvard investigation in Conan Doyle, *History of Spiritualism,* 1:211–29; and the cynical overview of the brothers' career in Harry Houdini, *A Magician among the Spirits* (New York: Harper and Brothers, 1924).

25: "How often has 'Science' killed off": William James's discussion of "spook philosophy" comes from "Confidences of a Psychical Researcher," first printed in the *American Magazine* in Oct. 1909. It is widely reprinted; for purposes of citation, I used Murphy and Ballou, *William James and Psychical Research,* 312.

26: "bloody howl of the Civil War": The discussion of James's family during the Civil War and the long-term effect on personal relationships and attitudes, even toward war, are beautifully told in Menand, *The Metaphysical Club,* 73–77 and 146–48.

27: "Now, don't, sir! Don't expose me!": "Mr. Sludge the Medium," in Robert Browning, *Dramatis Personae* (London: Chapman & Hall, 1864).

28: "such things are so indeed": Elizabeth Barrett Browning's letter to her sister is printed in Dingwall, *Some Human Oddities,* 128.

28: "a nauseating example": Ira Davenport's account of their problems in England, as well as an assessment of their career and the tricks involved in Houdini, *Magician among the Spirits.*

29: "an ingenious wire dummy": Other tricks in this selection are catalogued in Isaac Funk, *The Widow's Mite and Other Psychic Phenomena* (New York: Funk & Wagnall's, 1905); John Nevil Maskelyne, *Modern Spiritualism* (London: Frederick Warne, 1875); and David Abbott, *Behind the Scenes with the Medium* (Chicago: New Open Court, 1908).

30: "Stranger than Fiction": This story appeared in the August 1860 issue of the *Cornhill Magazine* and was written by an Irish journalist named Robert Bell.

30: "you would hold a different opinion": Thackeray's defense of spiritualism is cited in Inglis, *Natural and Supernatural,* 231.

2. A Spirit of Unbelief

33: famous (or infamous) book *On the Origin of Species:* As a graduate student at the University of Wisconsin, in the early 1980s, I took a history of science class, Darwin and the History of Biology, which stands in my memory as one of the best and most influential classes I ever attended. In it, we read three progressive editions of *On the Origin of Species,* through which we tracked Darwin's efforts to wrestle with and respond to the flood of attacks—both personal and scientific—which inundated him following publication of the first edition. I still have my favorite book from that class, which is a facsimile of the first edition, published by Harvard University Press in 1979. The first edition itself was published in London by John Murray, Albemarle Street, in Nov. 1859, and its full title was *On the Origin of Species by Natural Selection; or, The Preservation of Favoured Races in the Struggle for Life.*

I found two books and one dissertation helpful in thinking about the underlying tensions between science and religion at this point of time: Owen Chadwick, *The Secularization of the European Mind in the 19th Century* (Cambridge, England: Cambridge University Press, 1975); Frank Miller Turner, *Between Science and Religion: The Reaction to Scientific Naturalism in Late Victorian England* (New Haven, Conn.: Yale University Press, 1974); and John James Cerullo, "The Secularization of the Soul: Psychical Research in Britain, 1882–1920" (Ph.D. diss., University of Pennsylvania, 1980).

33: One author who proposed that the universe might have developed: The anonymous author was a Scottish journalist named Robert Chambers; the church's extremely angry reaction to his 1848 book *Vestiges of the Natural History of Creation* anticipated reaction to Darwin's work (as Darwin himself noted).

34: Alfred Russel Wallace: Wallace gets a very sympathetic portrait from Arthur Conan Doyle in *History of Spiritualism;* a complex portrayal in Turner, *Between Science and Religion,* 68–99; and a fairly critical analysis in Oppenheim, *Other World,* which includes some discussion of Charles Darwin's reactions to Wallace's spiritualist ventures.

Wallace describes his early investigations—and their implications—in a series of writings, including an 1866 pamphlet, *The Scientific Aspect of the Supernatural,* and a letter to the editor of the *Times* of London, Jan. 4, 1873, titled "Spiritualism and Science." Links to these and others of Wallace's writings, including his work on natural selection, can be found on Charles Smith's outstanding Alfred Russel Wallace page, www.wku.edu/~smithch/.

40: "I feel convinced": A. S. Sidgwick and E. M. S. Sidgwick, *Henry Sidgwick: A Memoir* (London: MacMillan, 1906), 187–88.

40: Henry Sidgwick: Sidgwick's letters and the biographical details of this section come from Sidgwick and Sidgwick, *Henry Sidgwick.* The original papers are archived at the Bodleian Library of Oxford University.

41: "When I found out how selfish": Sidgwick and Sidgwick, *Henry Sidgwick,* 271–72.

42: the son of a well-to-do Yorkshire clergyman: Myers writes about his life and his place in *Fragments of Inner Life: An Autobiographical Sketch* (privately printed, 1893; reprint, Society for Psychical Research, 1961).

44: "It may all be true": Huxley to Wallace, cited in James Marchant, *Alfred Russel Wallace: Letters and Reminiscences* (New York: Harper & Bros., 1916), 418; the book also details Wallace's early investigations of spiritualists and efforts to interest his colleagues. Huxley's second letter is quoted in *Report on Spiritualism, of the Committee of the London Dialectical Society,* Longman, Green & Co., London, 1871.

45: William Crookes: Crookes and his psychical research can be found in all good histories of the movement. I also found useful M. R. Barrington, ed., *Crookes and the Spirit World* (New York: Taplinger, 1972). The detail on thallium poisoning comes from Renee Haynes, *Society for Psychical Research, 1882–1982: A History* (London: MacDonald and Company, 1982), 179–81. Details of his early experiments are given in his "Spiritualism Viewed by the Light of Modern Science," *Quarterly Journal of Science,* July 1870; and "Notes of an Enquiry into the Phenomena called Spiritual During the Years 1870 through 1873," *Quarterly Journal of Science,* Jan. 1874. Crookes's controversial first report on D. D. Home was titled "Experimental Investigation of a New Force" and was in January 1871; his second (also referenced at the start of chapter 3) was done as the 1870–73 overview. Selected text from these papers and many others by psychical researchers can be found at the International Survivalist Society Web site: www.survivalafterdeath.org/home.htm.

45: One of the first mediums: Crookes's meeting with the planchette-wielding medium is reviewed in the *Encyclopedia for Psychical Research,* 199–201.

48: "We speak advisedly": Podmore, *Mediums of the Nineteenth Century,* 1:150–52, cites the article in *Quarterly Review,* Oct. 1871, and the fact that it was written anonymously by the famed physiologist W. B. Carpenter. Podmore, himself a notable psychic skeptic, characterizes Carpenter's attack on Crookes and Varley as "impaired by extraordinary egotism and malevolence." Crookes lodged a formal complaint against Carpenter with the Council of the Royal Society, forcing it to pass a resolution acknowledging that the statements in the *Review* had been inaccurate. But the incident did Carpenter no harm; the following year he was elected president of the British Association for the Advancement of Science.

49: "a December evening in 1871": Myers's thoughts are described in Myers, *Fragments of Inner Life;* and in Myers's obituary of Sidgwick in *Proceedings of the Society for Psychical Research* 15 (1901): 452–62.

3. Lights and Shadows

52: "a much perplexed man": Charles Darwin, *The Life and Letters of Charles Darwin* (New York: D. Appleton, 1888).

53: "worthless residuum of Spiritualism": The conclusion of Crookes, "Spiritualism Viewed by the Light of Modern Science."

54: "The impregnable position of science": John Tyndall, "Address Delivered before the British Association Assembled at Belfast, with Additions" (London: Longmans, Green, 1874). The text can be found at: http://www.victorianweb .org/science/science_texts/belfast.html.

55: "his warmest sympathies but no more": Gurney's first refusal to join in ghost hunting is cited in a letter from Sidgwick to Myers, from Sidgwick and Sidgwick, *Henry Sidgwick,* 288. Edmund Gurney has long intrigued historians. There are two books that focus on him particularly. Gordon Epperson, *The Mind of Edmund Gurney* (London: Fairleigh Dickinson University Press, 1997), offers a very kind portrait. Trevor H. Hall, in *The Strange Case of Edmund Gurney* (London: Gerald Duckworth, 1964), portrays the man as extraordinarily unstable. A balanced picture is given in Gauld, *Founders of Psychical Research.*

55: "those grand mysterious phenomena": A. R. Wallace, "A Defense of Modern Spiritualism," *Fortnightly Review* 15 (1874).

56: "mystery of the Universe": Gurney to W J, Sept. 23, 1883, Houghton.

56: their first serious investigation: The Newcastle investigations are described in Sidgwick and Sidgwick, *Henry Sidgwick;* and Gauld, *Founders of Psychical Research,* 107–14.

56: Sidgwick's attraction to Nora Balfour: The courtship and eventual marriage of Henry Sidgwick and Eleanor Balfour is based on accounts in Sidgwick and Sidgwick, *Henry Sidgwick,* 118, 301–306; and Ethel Sidgwick, *Mrs. Henry Sidgwick* (London: Sidgwick and Jackson, 1938), 48–50.

59: "prepared to be converted": Letter from Rayleigh to his mother, written after talking with Crookes and attending a sitting with Kate Fox-Jencken, cited in

Oppenheim, *Other World,* 331. Rayleigh's overall position on psychical research is outlined in Haynes, *Society for Psychical Research,* 198–99.

59: "Everything is always better": Letter to Myers in Sidgwick and Sidgwick, *Henry Sidgwick,* 301; Rayleigh letter to Sidgwick in Sidgwick, *Mrs. Henry Sidgwick,* 50.

59: new fad of "apports": Wallace wrote up his report on apports in the *Spiritualist,* Feb. 1, 1867.

60: a very pretty new medium: Among the many profiles of Anna Eva Fay, my favorite is in Eric J. Dingwall, *The Critic's Dilemma* (privately printed, 1966), 40–49, which does a terrific job of capturing her as a cheat and a complete charmer. The investigations of the Sidgwick group and interactions with William Crookes are recounted in Gauld, *Founders of Psychical Research;* Podmore, *Mediums of the Nineteenth Century,* 2:85, and Sidgwick and Sidgwick, *Henry Sidgwick,* 294.

61: Home took a fierce stand: D. D. Home, *Lights and Shadows of Spiritualism,* written in 1876 and published in 1877 by Virtue & Co. of London. In it, Home writes of receiving dismayed correspondence from Crookes and his wife and from leading spiritualists, one of whom wrote: "Astonishment! Surprise! Marvel! Have the heavens fallen upon you, Mr. Home, and crushed out your humanity? . . . I cannot think of a more ungracious, ill-repaying task than exposing the faults of others" (184).

62: His reports on Florence Cook were so smitten: Of the many accounts of William Crookes and Florence Cook, I found the most illuminating to be the "Florence Eliza Cook" chapter in Dingwall, *The Critic's Dilemma;* see also Podmore's ruthless exposé in *Mediums of the Nineteenth Century,* vol. 2, 97–103; and Pearsall, *The Table Rappers,* 93–100. The damage caused to Crookes's career, and his belated recognition of the risks of spending time with vindictive street mediums, are well covered in Inglis, *Natural and Supernatural,* which quotes Crookes's plaintive letter to Home on p. 276.

63: William Fletcher Barrett: Barrett's story appears in all of the histories of psychical research I have already referenced.

63: "If you can help me": The 1871 letter Barrett received from William Crookes is reprinted in E. E. Fournier D'Albe, *The Life of Sir William Crookes* (London: T. Fisher Unwin, 1923), 199.

64: "On Some Phenomena Associated with Abnormal Conditions of Mind": The paper rejected for publication by the BAAS was first printed in the *Spiritualist*

Newspaper 9 (Sept. 22, 1876): 85–94. A revised version appeared in the *Proceedings of the Society for Psychical Research* 1 (1882): 238–44.

65: "My opportunities have not been so good": Rayleigh's comments supporting Barrett's mind-reading studies to the British Association for the Advancement of Science are cited in Shepard, *Encyclopedia of Occultism and Parapsychology*, vol. 1, 770.

67: "a dreamy mystical face": The newspaper description of Slade's physical appearance is cited in Podmore, *Mediums of the Nineteenth Century*, 2:87.

67: insisted that Slade be prosecuted: Lankester's investigation of Slade and the resulting trial are detailed in all good histories of paranormal studies. Most of them also deal with the aftermath of his trial in England, which I did not. After fleeing to the Continent, Slade became the subject of some rather famous investigations by the German physicist and astronomer, Johann Zollner, of the University of Leipzig, who was then investigating mathematical notions of a fourth dimension in space. He ran a series of experiments with Slade, seeking to determine if the medium's apparent ability to levitate and transport objects was due to a talent for accessing the fourth dimension. Slade apparently performed brilliantly in these experiments, but the result was a broadside scientific attack on Zollner's reputation, including suggestions that he was suffering from senility, although the scientist was only in his early forties at the time.

Henry Slade's story is summarized in Shepard, *Encyclopedia of Occultism and Parapsychology*, vol. 1, 838–40, which describes the last days of his life with some self-righteousness: "He fell victim to the drink habit, his moral standing was far from high and he sank lower and lower. He died penniless and in mental decrepitude in a Michigan sanitorium in 1905."

68: a woman he could not have: Frederic Myers's love affair with Annie Marshall dominates his autobiographical sketch, *Fragments of Inner Life*, and is recounted in all histories of the psychical research movement, in the most detail in Gauld, *Founders of Psychical Research*, 116–24.

69: Myers caught the dark scent of the occult: His sittings with the Parisian mediums are described in a letter to Henry Sidgwick on August 16, 1877, archived at the Society for Psychical Research, London, and reprinted in Gauld (1968).

70: "hints that Gurney's pretty wife . . . had married for money": The most critical view of the marriage between Edmund Gurney and Sara Kate Sibley can be found in E. A. Sheppard, *Henry James and "The Turn of the Screw"* (Auckland, New Zealand: Auckland University Press, 1974), 128–33, which speculates that the troubled relationship formed the background for one of James's later stories.

70: "You can live without me": Gauld, *Founders of Psychical Research,* 134–36, cites Evie Myers's letters. The story of Myers's courtship of Eveleen Tennant (and the implication that Annie Marshall overshadowed it from the beginning) is given in Trevor Hall, "The Mourning Years of F. W. H. Myers," *Tomorrow* 12 (1964).

71: William Barrett published his second report: Barrett's second report on thought transference was first summarized in *Nature* in July 1881 under the title "Mind-Reading versus Muscle-Reading." He coauthored a longer and more detailed article, based on those results, with Edmund Gurney and Frederic Myers, published in the *Proceedings of the Society for Psychical Research* 1 (1882), the same issue in which his first report appeared.

72: The British Society for Psychical Research formally convened: Founding of SPR and Sidgwick's presidential address is reproduced in Sidgwick and Sidgwick, *Henry Sidgwick,* 360–64.

73: A dream that had haunted him for twenty-four years: The story of Clemens's precognitive dream of his brother's death is well known. I based my version on several accounts, including the very straightforward one given at the Web site about.com, given in its overview of the paranormal: http://paranormal.about .com/od/othermystics/a/aa090604_2.htm.

4. Metaphysics and Metatrousers

75: "the inside of a coal mine": WJ to Charles Renouvier, Dec. 16, 1882, Houghton.

78: "the old man stubbornly turned his face to the wall": The description of James's father's death is based on multiple accounts, including those in Simon, *Genuine Reality;* Feinstein, *Becoming William James;* and Jean Strouse, *Alice James* (Boston: Houghton Mifflin, 1980).

78: unexpected invitation from Edward Gurney: Gurney's invitation to WJ, Dec. 13, 1882, Houghton.

79: "one of the first rate minds of the time": WJ to wife, Dec. 18, 1882; "I doubt its compatibility": Gurney to WJ, Sept. 23, 1883; both in Houghton.

80: "No matter where you open its pages": "What Psychical Research Has Accomplished," *Forum,* Aug. 13, 1892, 727–42.

80: "I have been tremendously busy": Gurney to WJ, Feb. 18, 1884, Houghton.

81: "belief in new physical facts": WJ to Thomas Davidson, Mar. 30, 1884, Houghton.

81: "Of all the senseless babble": From William James, "Are We Automata?" *Mind* 4 (1879): 1–22.

82: A few religious leaders: The religious response to the challenges of science is well told in Ronald Numbers, *Darwinism Comes to America* (Cambridge, Mass.: Harvard University Press, 1998).

82: "I confess I rather despair": WJ to Davison, Mar. 30, 1884, Houghton.

82: "remarkable phenomena": William Barrett, "The Prospects of Psychical Research in America," *Journal of the Society for Psychical Research* 6 (1884).

83: "paralyze the phenomena": Henry Sidgwick's belief that he paralyzed spiritual phenomena in Sidgwick and Sidgwick, *Henry Sidgwick,* 284; his analysis of psychical researchers, including himself, can be found on pp. 387–88 of the same work.

83: "a cheerful cynic named Richard Hodgson": For my portrayal of Richard Hodgson, I draw largely on a lengthy collection of intimate letters that Hodgson wrote to his friend Jimmy Hackett, describing his days, thoughts, hopes, and plans in extraordinary detail. These letters are archived at the American Society for Psychical Research. Two other excellent sources of information are Alex Baird, *The Life of Richard Hodgson* (London: Psychic Press, 1949); and Arthur Berger, *Lives and Letters in American Parapsychology: A Biographical History, 1850–1987* (Jefferson, N.C.: McFarland, 1988). Berger's chapter on Hodgson (11–33) is meticulously researched and objectively told; Baird's book is something of a hagiography. All Hodgson quotes in this section are from Hodgson's letters to Hackett, including the Wordsworth verse and the evolution doggerel.

85: "Blavatsky carried such a reputation": Helena Petrovna Blavatsky's remarkable rise is described in "Madame Blavatsky's Baboon: A Spirited Story of the Psychic and the Colonel," *Smithsonian,* May 1995, 110–28, and is one of the more famous (and entertaining) stories in the history of the psychical research movement. Berger, *Lives and Letters in American Parapsychology,* focuses particularly on Hodgson's exposé, including the Australian's suspicions that the so-called psychic might actually be a Russian spy. Conan Doyle, *The History of Spiritualism* (1:260–62), offers a more sympathetic version of her story.

86: "She is a genuine being": Sidgwick and Sidgwick, *Henry Sidgwick,* 384–85.

86: "The evidence published by the English society": The first circular of the ASPR, with its organizing principles, can be found in Berkhardt and Bowers,

Essays in Psychical Research, 5-8. Founding of ASPR is described in a variety of sources, including Moore, *In Search of White Crows,* 142–44; Simon, *Genuine Reality,* 190–92; and Berger, *Lives and Letters in American Psychology,* 8–11. James comments on the importance of a strong science component are in a letter to philosopher Thomas Davidson on February 1, 1885, as well as his prediction that Newcomb would probably "carry the others" with him.

87: dismissed Barrett's work on mind reading: Newcomb wrote first about his problems with the British experiments in "Psychic Force," *Science,* Oct. 17, 1884; he expanded his criticisms in *Science,* Jan. 29, 1886, in an essay based on his ASPR presidential address of early 1885.

88: "It is worrying to think": Gurney's reply appeared in *Science* on Dec. 5, 1884.

89: "considerably off the rails": Gurney to WJ, July 31, 1885, Houghton.

89: coined a new name: The SPR's early experiments on telepathy, and Myers's invention of the word, are described in many venues, including Funk, *Widow's Mite,* 294–309; Rosalind Heywood, *The Sixth Sense* (London: Pan Books, 1971), 39–42; and G. N. M. Tyrrell, *The Personality of Man* (West Drafton, Middlesex, England: Pelican Books, 1948).

90: "To brand as dupes": James, having done his drawing experiments, wrote *Science* on January 30, 1886, to complain. He followed that up with a letter to Newcomb on February 12, 1886, with samples of his own drawings, and wrote to the astronomer again that summer, to tell him that he was wrong about thought transference.

91: "The whole thing is a fraud": Hodgson correspondence to Jimmy Hackett, March 19, 1885, ASPR.

92: "this perpetual association": Gurney to WJ, Mar. 31, 1885, Houghton.

92: a country squire in eastern England: The squire's story comes from Edmund Gurney, *Phantasms of the Living* (London: Trubner, 1886). I worked both from a facsimile copy of the original edition and from a compact revised edition, edited and abridged by Nora Sidgwick (1918; reprint, Hyde Park, N.Y.: University Books, 1962). The squire's story, as retold by me, is based on a series of letters on pages 163–66 of the original edition.

94: "absolutely *reek* of candour": Mar. 31, 1885, after Newcomb made his presidential address.

94: analysis of ghost stories: Nora Sidgwick, "Notes on the Evidence, Collected by the Society, for Phantasms of the Dead," *Proceedings of the Society for Psychical Research* 3 (1885): 69–150.

95: His third son, Herman: On July 11, 1885, William James wrote a letter to his aunt, Kate Walsh, about the burial of his son, Herman, which was among many grieving notes he wrote to friends and relatives. His visit to the old house is described in an August 28, 1885 letter to his wife. Simon, *Genuine Reality* (196–200), discusses both the child's death and the way that grief led the Jameses to meet with the Boston medium, Leonora Piper.

97: Leonora Evelina Piper: Among the countless sources for the life of Leonora Piper, I would like to reference: Alta Piper, *The Life and Work of Mrs. Piper* (London: Kegan Paul, 1929); Anne Manning Robbins, *Past and Present with Mrs. Piper* (New York: Henry Holt, 1922); Podmore, *Mediums of the Nineteenth Century,* 2:308–29; Amy Tanner, *Studies in Spiritism* (New York: D. Appleton, 1910), 9–46; and Berkhardt and Bowers, *Essays in Psychical Research,* from the notes section, pp. 394–400

99: "I remember playing the *esprit fort*": William James's account of his first meeting and subsequent interest in Leonora Piper was published as "A Record of Observations of Certain Phenomena of Trance," *Proceedings of the Society for Psychical Research* 6 (1890): 651–59. The report was read to the society by his brother, Henry James Jr.

100: "People who fly into rages": The Sidgwick discussion of spiritualists from Sidgwick and Sidgwick, *Henry Sidgwick,* 425, and Sidgwick, *Mrs. Henry Sidgwick,* p. 99; R. Hodgson and S. J. Davey, "The Possibilities of Malobservation and Lapse of Memory from a Practical Point of View," *Proceedings of the Society for Psychical Research* 4 (1887): 381–495; Wallace's response cited in Gauld, *Founders of Psychical Research.*

102: "'he says his name is John'": Minot Savage's accounts of his early investigations of Leonora Piper are detailed in Minot J. Savage, *Can Telepathy Explain?* (New York: G. P. Putnam's Sons, 1902).

104: James and Savage collaborated on a formal report: "Report of the Committee on Mediumistic Phenomena," *Proceedings of the American Society for Psychical Research* 1 (1882): 102–6.

5. Infinite Rationality

105: "I was only waiting for breath": Gurney to WJ, Apr. 16, 1886, Houghton.

106: an essay on the fleeting brilliance of nitrous oxide: William James, "Subjective Effects of Nitrous Oxide," *Mind* 7 (1882).

106: both experimented with hashish: Hodgson describes his and Myers's use of hashish in a letter to Jimmy Hackett on Mar. 4, 1886, ASPR.

106: "On Cocaine": *Journal of Substance Abuse Treatment* 1, no. 3 (1984): 206–17.

107: Frank Podmore: An excellent short biography of Frank Podmore can be found in the psychical research scholar Eric J. Dingwall's introduction (v–xxii) to Podmore, *Mediums of the Nineteenth Century.*

107: asked Podmore and Myers to investigate: Podmore's incisive analysis of D. D. Home in Podmore, *Mediums of the Nineteenth Century,* also includes background on his investigation with Myers and their differences of opinion.

109: "I was palpably aiming": Sidgwick's journal for Jan. 4, 1885, cited in Gauld, *Founders of Psychical Research,* 161; archived at the Bodleian Library, Oxford University.

109: The provocative question: Myers's discussion of religion and psychical research in the introduction to Gurney, *Phantasms,* x–xii.

110: "is it wise to say it?": Journal entry for June 28, 1885, in Sidgwick and Sidgwick, *Henry Sidgwick,* 415.

111: "Planchette is simply nowhere": "The New 'Planchette': A Mysterious Talking Board and Table over Which Northern Ohio Is Agitated," *New York Tribune,* Mar. 28, 1886.

112: "Attention to such gruesome tales": Description of Gurney's house and meeting with Lodge from Oliver Lodge, *Past Years: An Autobiography* (New York: Charles Scribner's Sons, 1932), 270–71.

112: each ghost story gained power: All "ghost stories" are retellings of those in Gurney, *Phantasms.*

116: "The Bloodthirsty Bluebells": Sidgwick and Sidgwick, *Henry Sidgwick,* 408.

118: "fail to reveal any sensibility for a magnetic field": Joseph Jastrow and George F. H. Nuttall, "On the Existence of a Magnetic Sense," *Proceedings of the American Society for Psychical Research* 1 (1882): 116–26.

118: a pattern of insanity: Suggestions of mental aberration form the central thesis of Josiah Royce, "Report of the Committee on Apparitions and Haunted Houses," *Proceedings of the American Society for Psychical Research* 1 (1882): 224.

119: "I think your constant allegation of fraud": Two letters from Alfred Russel Wallace to William James, written Dec. 21, 1886, and Dec. 23, 1886, enumerate his complaints against psychical research. The letters were prompted by

James's work in publicly exposing a Boston medium, Hannah Ross, whom Wallace admired.

120: "select the weak stories": Sidgwick's journal for Aug. 22, 1885, in Sidgwick and Sidgwick, *Henry Sidgwick.*

120 "monotonous assortment": Gurney to WJ on Apr. 16, 1886, Houghton.

120: "ruthless hand of science": Myers's commentary on science in psychical research from his introduction to Gurney, *Phantasms,* xxiii–xxvii.

121: "theories on which to build": Myers's list of theories in ibid., xxxii–xxxiii.

121: "ploughing through some strange ocean": Ibid., xxxv–xxxvi.

122: Powdered nutmeg: Results from Gurney's taste testing, ibid., 46–49.

125: "The question for us now": . . . and discussion of census: Ibid., 376–89.

127: "a bit of land": WJ to his wife, Nov. 11, 1886, Houghton.

127: "mind-cure doctress": James's doctress and his stand against Massachusetts legislation to eliminate alternative healers are discussed in the introduction to Murphy and Ballou, *William James and Psychical Research,* 8–12.

127: the bane of the American Medical Association: The early AMA position and Mark Twain's response are described in Patrick K. Ober, M.D., "Mark Twain's Criticism of Medicine in the United States," *Annals of Internal Medicine* 126, no. 2 (Jan. 15, 1997): 157–63.

129: "This is a most extraordinary work": William James, "Review of *Phantasms of the Living,* by Edmund Gurney et al.," *Science* 9 (Jan. 7, 1887): 18–20.

129: research community had taken notice: Scientific response to *Phantasms,* with an emphasis on the "tone of superior wisdom" taken by Charles Peirce, in Gauld, *Founders of Psychical Research,* 171–74.

6. All Ye Who Enter Here

131: "der bestirnte Himmel": Immanuel Kant, *Critique of Practical Reason,* reprinted, Cambridge University Press, Cambridge, 1997.

131: "Sidgwick . . . as a young philosopher": For the context for Sidgwick's place in philosophy, see Barton Schultz, "Henry Sidgwick," in *The Stanford Encyclopedia of Philosophy* (winter 2004 edition), ed. Edward N. Zalta. http://plato.stanford.edu/archives/win2004/entries/sidgwick/.

132: "a Sovereign will": Sidgwick and Sidgwick, *Henry Sidgwick,* 373–74.

133: "He combines the powers": Gurney to WJ, Jan. 16, 1887, Houghton.

134: "Well, I am happy enough": Hodgson to Hackett, July 18, 1887, and Nov. 14, 1887, ASPR.

135: "My cousin Fred": Hodgson, "A Record of Observations of Certain Phenomena of Trance," *Proceedings of the Society of Psychical Research* 8 (1892): 60–67.

136: "a dissipated looking wreck": Joseph Rinn, *Sixty Years of Psychical Research* (New York: Truth Seeker, 1960), 76.

136: a new scientific commission: The Report of the Seybert Commission on Spiritualism (J. B. Lippincott Company, Philadelphia, 1920), 32–45.

137: rare united reaction: The spiritualist response is described in Funk, *Widow's Mite,* 81–83, 117–19.

138: "Professor James and his ilk": James to the editor of the *Banner of Light,* Feb. 10, 1887; *Banner* printed his letter and the editors' response in the same issue.

138 : "the acumen of Hodgson": Sidgwick and Sidgwick, *Henry Sidgwick,* 465.

139: Alice and Mary were caught signaling: The Creery sisters' cheating is described in Hall, *Strange Case of Edmund Gurney,* 55–61; Oppenheim, *Other World,* 359–60.

140: "If only I could form the least conception": Sidgwick, Mar. 30, 1887, cited in Sidgwick and Sidgwick, *Henry Sidgwick,* 374.

141: "In one sense, I am sacrificing myself": Hodgson to Hackett, Jan. 29, 1888, ASPR.

142: It took all James's diplomatic skills: James as diplomat, in Piper, *Life and Work,* 43–45.

142: "You idiot!": Piper, *Life and Work,* 62; Hodgson's other precautions are described on pp. 62–65.

143: "It is a mercy that Hodgson exists": Gurney to WJ, May 20, 1888, Houghton.

144: Gurney's death: Obituary in the *Atheneum,* June 30, 1888, 827; account of inquest in Hall, *Strange Case of Edmund Gurney,* based on accounts from the *Sussex Daily News, Brighton Examiner, Brighton Gazette, Brighton and Hove Herald,* June 26, 1888; and Epperson, *Mind of Edmund Gurney,* 137–53. Three books also discuss Gurney's suicide, with Hall raising the possibility that he was demoralized by cheating in the SPR's telepathy experiments. Epperson contradicts this point of view emphatically, as does Gauld, *Founders of Psychical Research,* 173–82; both of these authors prefer the idea of either accident or a suicide related to his manic-depressive personality.

144: "It seems one of Death's stupidest strokes": WJ to Henry James; cited in Sheppard, *Henry James,* 122.

146: "fury of this hunt after ghosts": George Croom Robertson, editor of *Mind,* to WJ, Sept. 9, 1888, Houghton.

146: "whispers . . . about his marriage": Gauld, *Founders of Psychical Research,* touches briefly on the sympathy for Kate Gurney's loneliness; the problems of the marriage are more thoroughly discussed in Sheppard, *Henry James* (123–31), including her refusal to donate to memorial fund, and encounter with Henry James Jr. in Paris.

147: Charles Richet: Richet's personality described in Lodge, *Past Years,* 291–92.

147: French researchers approached hypnosis: French scientists' work with hypnotism is reviewed in Inglis, *Natural and Supernatural,* 338–53, and Epperson, *Mind of Edmund Gurney,* 65–67; it is also covered well by Richet himself: Charles Richet, *Thirty Years of Psychical Research* (New York: Macmillan, 1923), 120–30.

149: a temperamental Italian peasant woman: Eusapia Palladino appears in every history of spiritualism published. For biographies contemporary to her time, I used Hereward Carrington, *Eusapia Palladino and her Phenomena* (New York: B. W. Dodge, 1909); and Theodore Flournoy, *Spiritism and Psychology* (New York: Harper and Brothers, 1911), 242–60.

150: Maggie Fox Kane sold a confession: Maggie Fox Kane's confession from Funk, *Widow's Mite,* 240–42; "Mrs. Margaret Fox-Kane's Confession," *New York World,* Oct. 21, 1888; Rinn, *Sixty Years,* 55–56, 76–80, discusses journalist's planned book.

151: Census of Hallucinations: *International Congress of Experimental Psychology: Instructions to the Person Undertaking to Collect Answers to the Question on the Other Side* (1889), from the archives of the American Society for Psychical Research; letter from William James seeking publicity for the census, "To the Editor of the American Journal for Psychology," *American Journal of Psychology* 3 (Apr. 3, 1890): 292.

152: "deplorably hasty": Myers, letter to James on maintaining the American branch of the Society for Psychical Research, Dec. 12, 1888, Houghton.

152: Miss Mary A. T. Sitting: Hodgson, "Certain Phenomena of Trance," 67–167.

154: "I am quite thick now with Sidgwick": WJ to his wife, Aug. 7, 1889, Houghton.

155: "A curious chapter": Richet, *Thirty Years,* 26–27. On the instability of mediums, see also Funk, *Widow's Mite,* 105–111; and Barrett, *Threshold of the Unseen,* 123–24.

156: "I'm so cold": Hodgson, "Certain Phenomena of Trance," 92–95.

7. The Principles of Psychology

157: the Metaphysical Society: R. H. Hutton, "The Metaphysical Society: A Reminiscence" *The Nineteenth Century (1885)* posted on The Huxley File website, Created by Charles Blinderman, Professor of English and Adjunct Professor of Biology, and David Joyce, Professor of Mathematics and Computer Science, Clark University http://aleph0.clarku.edu/huxley/.

158: *Christianity and Agnosticism:* Huxley's essay "Agnosticism" first appeared in *The Nineteenth Century* in Feb. 1889 and was reprinted in a collection of essays on the subject, *Christianity and Agnosticism* (London: D. Appleton, 1889).

159: "Mrs. Piper . . . flatly refused": For Mrs. Piper's refusal and eventual agreement and Myers's letter, see Piper, *Life and Work,* 47–53.

160: "Don't ky, Alta": Ibid., 50.

161: "Why Mrs. Piper": Ibid., 53

163: fanatical preparations: For the sittings with Myers and Lodge and descriptions of house and hospitality, see ibid., 54–61; Oliver Lodge, *The Survival of Man* (London: Methuen, 1909), 458–59; Funk, *Widow's Mite,* 240–42; and Barrett, *Threshold of the Unseen,* 26–30.

167: "I am filled with confusion": Shepard, *Encyclopedia of Occultism and Parapsychology,* 1:522. For background on Eusapia Palladino, see also Carrington, *Eusapia Palladino;* and Flournoy, *Spiritism and Psychology,* 242–60.

168: *The Principles of Psychology:* William James, *The Principles of Psychology* (Boston: Henry Holt, 1890).

168: "No one . . . ever had a simple sensation": Ibid., 1:224–46.

168: "Objects of rage, love": Ibid., 2:449.

169: "a man has as many social selves": Ibid., 1:300–302.

169: "Dear Bill": Holmes to WJ, Nov. 10, 1890, Houghton.

170: "your magnificent book": Hall to WJ, Oct. 14, 1890, Houghton.

170: "It is literature": Cited in Ernest R. Hilgard, *Psychology in America: A Historical Survey* (New York: Harcourt Brace Jovanovich, 1987), in an excellent section titled "James as an Unsystematic Psychologist," 48–56, which gives an overview of scientific reception, including further comments from Hall and from William Wundt.

170: "big and good book": Myers to James, Jan. 12, 1891; James response, Jan. 20, 1891, Houghton.

170: "the most incorrigibly and exasperatingly critical":"What Psychical Research has Accomplished," *Scribner's,* 1890, revised and expanded for *Forum,* Aug. 13, 1892, 727–47.

171: the "nervous weakness": WJ to his sister, July 6, 1891, Houghton.

172: "when I am gone": Alice to WJ, July 30, 1891, Houghton.

172: "She talks death": WJ to his wife, Sept. 25, 1891, Houghton.

172: "the dreadful Mrs. Piper": Alice James, *Diary of Alice James* (New York: Mead, 1964), 231.

172: Mark Twain published a personal endorsement: Mark Twain, "Mental Telegraphy: A Manuscript with a History," *Harper's New Monthly Magazine,* v. 84 (Dec. 1891): 95–104.

174: inspired response appeared in *Scribner's:* "The Logic of Mental Telegraphy", Joseph Jastrow, *Scribner's,* Jan. 1892. For more perspective on the increasing hostility of scientists like Jastrow toward psychical research, see Joseph Jastrow, "The Problems of Psychical Research," *Harper's Magazine* 79 (June 1889).

175: insider exposé of spiritualism: A. Medium, *Revelations of a Spirit Medium* (St. Paul, Minn.: Farrington, 1891). This was one of two books considered by Richard Hodgson to be the best insider guides to the medium trade. The other was John Truesdell, *The Bottom Facts Concerning the Science of Spiritualism: Derived from Careful Investigations Covering a Period of Twenty-five Years* (New York: G. W. Dillingham, 1892).

177: Census of Hallucinations: For background on the census, see Sidgwick, *Mrs. Henry Sidgwick,* 121–25; and Sidgwick and Sidgwick, *Henry Sidgwick,* 501–2.

179: "Her neurotic temperament": WJ to Henry James, Mar. 6, 1892, Houghton.

179: "If you were here": Mar. 7, 1892, Houghton; cited in Simon, *Genuine Reality,* 241.

179: "God (or the unknowable)": WJ to Hodgson, May 15, 1892; letter on census, May 25, 1892, Houghton.

180: "No one is saying": Sidgwick's journal, May 2, 1892, in Sidgwick and Sidgwick, *Henry Sidgwick.*

181: first report on Mrs. Piper: Hodgson, "Certain Phenomena of Trance."

181: "Oh, how black": Piper, *Life and Work,* 67.

182: "Between the deaths": "The Report on the Census of Hallucinations," *Proceedings of the Society for Psychical Research* 10 (1894): 25–422; William James, "Review of the 'Report on the Census of Hallucinations,'" *Psychological Review,* Jan. 2, 1895, 69–75. Publication followed Sidgwick's 1892 report to the Congress of Experimental Psychology.

184: "I never believed": WJ to Henry Sidgwick, July 11, 1896; cited in Berkhardt and Bowers, *Essays in Psychical Research,* 74.

8. The Invention of Ectoplasm

185: among that season's many victims: The strange story of George Pellew is told in Piper, *Life and Work,* 77–79, 104–7; Baird, *Life of Richard Hodgson,* 65–73; Tanner, *Studies in Spiritism,* 26–27; and Richet, *Thirty Years,* among others.

189: "Your letter rec'd": WJ to Myers, Nov. 14, 1892, Houghton.

190: "So runs the world away!": William James, "Frederic Myers' Service to Psychology," *Proceedings of the Society for Psychical Research* 17 (1901): 13–23.

191: "The *first* reason": Myers to WJ, Oct. 10, 1893, Houghton.

191: "It seems to me you lack": Myers to WJ, Nov. 16, 1893, Houghton.

192: "James accepts": WJ to Myers, telegraphed acceptance, Dec. 17, 1893, Houghton.

192: scientific investigations of Eusapia Palladino: Richet, *Thirty Years,* 400–410, 454–58; Gauld, *Founders of Psychical Research;* and Everard Feilding, Wortley Baggally, and Hereward Carrington, "Report on a Series of Sittings with Eusapia Palladino," *Proceedings of the Society for Psychical Research* 23 (1909): 309–20.

194: Lodge remembered the journey: Account of the sittings on Ile Roubaud is taken from Lodge, *Past Years,* 292–306; see also *Journal of the Society for Psychical Research* 6 (1894): 350–57.

196: "crisis": Sidgwick to H. G. Dakyns, July 29, 1894; in Sidgwick and Sidgwick, *Henry Sidgwick.*

197: "greatest donkey of the age": WJ to Myers on Lombroso, Sept. 9, 1894, Houghton.

198: "You ask what I think": WJ to Lodge regarding caution in publishing, Oct. 4, 1894, Houghton.

199: She published his analysis: R. Hodgson, *Journal of the Society of Psychical Research* 7 (1895): 55–79.

201: they would test Eusapia on their terrain: Eusapia's sittings in Cambridge and response described in Sidgwick, *Mrs. Henry Sidgwick,* 138–40; Inglis, *Natural and Supernatural,* 387–88; and Podmore, *Mediums of the Nineteenth Century,* 198–203.

201: "Sidgwick has to *flirt* with her": Myers to WJ, Aug. 8, 1895, Houghton.

203: "Well, our countries": WJ to Myers, Jan. 1, 1895, Houghton.

204: "It has not been the practice": Sidgwick in *Journal of the Society for Psychical*

Research 7 (1895); "I fear the Eusapia business": WJ to Sidgwick, Nov. 8, 1895, Houghton.

204: "The Presidency of the Society": William James, "Address by the President," *Proceedings of the Society for Psychical Research* 12 (1896): 2–10.

205: "too hasty assumption": Haynes, *Society for Psychical Research,* 180.

9. The Unearthly Archive

209: "Do not allude to all this": Myers to WJ, Dec. 21, 1900, Houghton.

210: "I do not say that facts": To Lodge, Oct. 10, 1893; cited in Gauld, *Founders of Psychical Research.*

211: "Studies in Hysteria": Cerullo, "Secularization of the Soul," 133–34, 221.

211: compared the range of human consciousness to the light spectrum: Flournoy, *Spiritism and Psychology;* Moore, *In Search of White Crows,* 149–52.

212: "a narrowish intellect": WJ to Charles William Eliot, Feb. 21, 1899, Houghton.

213: letter to the *Psychological Review:* J. Cattell, "Psychical Research," *Psychological Review* 3 (1896): 582–83.

213: "snarling logicality": "The Will to Believe," address to the Philosophical Clubs of Yale and Brown Universities, *New World,* 1896; reprinted in Stephen Rowe, *The Vision of James* (London: Vega, 2001).

214: "About the narrowest minded": Barrett to Lodge, Oct. 14, 1893; quoted in Oppenheim, *Other World,* in section on dowsing, 362–64. The Divining-rod entry, Shepard, *Encyclopedia of Occultism and Parapsychology,* 250–52, includes a summary of Barrett's report.

218: second report on Leonora Piper: Richard Hodgson, "A Further Record of Observations of Certain Phenomena of Trance: Additional Report on Mrs. Piper," *Proceedings of the Society for Psychical Research* 13 (1898): 284–583.

220: "If Professor Sidgwick": For Hodgson's theories of communication from Barrett, see Barrett, *Threshold of the Unseen,* 243–49; for further discussion, see Frederic Myers, "On Some Fresh Facts Indicating Man's Survival of Death," *National Review,* Apr. 1898, 230–42.

221: "fired off an essay": James McKeen Cattell, "Mrs. Piper, the Medium," *Science,* Apr. 15, 1898. *Science,* May 6, 1898, published James's "Letter on Mrs. Piper, the Medium," along with Cattell's response; James's review of Hodgson's report was published in *Psychological Review,* July 1898, 420–24.

223: "art of unveiling fraud": Hodgson's report is discussed in the *Saturday Review,* July 16, 1898, 81.

224: "a cosmic record of sorts": James's discussion of the "cosmic consciousness" is best explained in one of his last works, "Confidences of a Psychical Researcher," published in *American Magazine* 69 (Oct. 1909): 580–89.

225: "eleven reported poltergeist cases": Podmore's poltergeist investigations and his belief that they helped explain Palladino are also discussed in "On Poltergeists," a chapter in his *Mediums of the Nineteenth Century.*

226: Theodore Flournoy: An excellent review of the relationship between William James and Theodore Flournoy can be found in the introduction to *The Letters of William James and Theodore Flournoy,* ed. Robert C. Le Clair (Madison: University of Wisconsin Press, 1966).

226: "every finger of": Flournoy to WJ, Dec. 11, 1898, Houghton.

227: "intentional and systematic fraud": Sidgwick cited in Gauld, *Founders of Psychical Research,* 241, based on a report in *Proceedings of the Society for Psychical Research* 35 (1938): 165.

227: "a grave moment for all of us": WJ to Lodge, Oct. 4, 1894, Houghton.

227: "The Canterville Ghost": In Oscar Wilde, *Lord Arthur Savile's Crime and Other Stories* (London: James Osgood, 1891).

228: "Turn of the Screw": Sheppard, *Henry James,* 141–43.

230: "a manifestation of persistent energy": *Proceedings of the Society of Psychical Research* 6 (1889): 15–16. Henry James's knowledge of the psychical research movement is discussed in Epperson, *Mind of Edmund Gurney,* 116–212; and Beidler, *Ghosts, Demons and Henry James.*

230: "perhaps among my audience": Crooke's presidential address to the British Association for the Advancement of Science, Oct. 1898, Bristol, England.

230: James Hervey Hyslop: The story of James Hyslop is drawn primarily from his unpublished biography in the archives of the American Society for Psychical Research, and all quotes about his childhood are from those documents. Hyslop describes his early experiments with Mrs. Piper in those documents and in nearly all the books he wrote later in life. Moore, *In Search of White Crows* (159–65), offers a concise biography; and Berger, *Lives and Letters in American Parapsychology* (35–65), also provides an excellent portrait of Hyslop.

232: "No scientifically-minded psychologist": Berkhardt and Bowers, *Essays in Psychical Research,* 167–79, reprints the letters exchanged by James and Titchener in *Science.* Titchener to WJ on psychical research, May 28, 1899, following two letters from WJ, May 6, 1899, and May 21, 1899.

234: Rosina Thompson: Thompson's mediumship is recounted in "On the Trance Phenomena of Mrs. Thompson," *Proceedings of the Society for Psychical Research* 17 (1902): 67–74, an article based on a talk given by Frederick Myers in July 1900. Additional information can be found in Richet, *Thirty Years,* 284–91; Gauld, *Founders of Psychical Research,* 268–74; Berger, *Lives and Letters in American Parapsychology,* 27–29; Podmore, *Mediums of the Nineteenth Century,* vol. 2, 357; and John Piddington, "On the Types of Phenomena Displayed in Mrs. Thompson's Trance," *Proceedings of the Society for Psychical Research* 18 (1904): 104–307.

235: "My first sittings": Myers to WJ, Oct. 14, 1899, Houghton.

10. A Prophecy of Death

239: "He called the response 'anaphylaxis'": Richet's work on anaphylaxis from the Nobel Prize for Medicine presentation speech by Professor C. Sundberg, vice-chairman of the Nobel Committee for Physiology or Medicine, Dec. 10, 1913. The text can be found at: http://nobelprize.org/medicine/laureates/1913/press.html.

240: "So low had this unfortunate woman sunk": Funk, *Widow's Mite,* 241.

241: "will was recovered": "Says Will Was Found through Spirit Medium; Mrs. Mellen Tells of Feat of One of the Fox Sisters," *New York Times,* Mar. 16, 1905.

241: *From India to the Planet Mars:* Theodore Flournoy, *From India to the Planet Mars: A Study of a Case of Somnambulism* (New York: Harper & Brothers, 1900).

243: "Your book has only one defect": WF to Flournoy, Jan. 1, 1900, Houghton.

243: "*tho* with extreme slowness": WF to Hodgson, Jan. 19, 1900, Houghton.

244: "We are having the D—l's own time": WJ to Henry James III, Feb. 23, 1900.

244: "Your mother is extremely rosy": WJ to Margaret Mary James, Mar. 10, 1900.

244: "My Dear Mother": Hyslop's questions home and the answers from his family are archived in his correspondence files at the American Society for Psychical Research, as are the letters between Hyslop and Hodgson.

248: found himself in a fight for survival: Hyslop's problems at Columbia began when he discussed his sittings with Mrs. Piper in conferences held at Cambridge in June 1899. It was from that meeting that reporters began writing that he planned to scientifically prove immortality. Although he sent denials to both *Science* and *Psychology Review,* his fellow faculty members were swayed by his continued public writings, repeating his belief in immortality, in national magazines; see, for example "Results of Psychical Research," *Harper's Magazine* 100 (Apr. 1900): 786–97.

248: "It would be pretty absurd": Hodgson's letter to Hyslop, Feb. 23, 1900. This and all correspondence that I cite comes from unpublished letters between Hyslop and Hodgson, ASPR.

249: "Life is very strange now": Sidgwick to Myers, May 24, 1900, in Sidgwick and Sidgwick, *Henry Sidgwick,* 587.

250: "Dear Mrs. Sidgwick": WJ to Nora Sidgwick, Sept. 1, 1900, Houghton.

250: "My brain power": WJ to Myers, Dec. 6, 1900.

251: "his subliminal is": WJ to John Piddington, Jan. 5, 1901.

251: "That intolerable babbler": WJ to Frances Morse, Jan. 4, 1901.

251: "I think of you": WJ to Myers, Dec. 8, 1901.

252: "His serenity": WJ to Nora Sidgwick, Jan. 20, 1901, Houghton.

252: "Is there going to be any difficulty": For the A control and other secrets, see WJ to Lodge, Mar. 16, 1901; reply from Lodge, Mar. 19, 1901. Evie Myers response, including demands for destruction of documents, summarized in Berger, *Lives and Letters in American Parapsychology,* 30. Letters from Evie to WJ, Mar. 24, 1902, and Apr. 17, 1902, Houghton. A version with all references to Annie Marshall deleted was published in 1904 as *Fragments of Prose and Poetry*; the argument, though, continued for years. In the summer of 1906, James was still writing to Evie Myers, refusing to destroy letters and documents or to return all copies of the unabridged autobiography.

254: "being absolutely fearless": Lodge to WJ, Mar. 19, 1901.

255: "I poured experimental telepathy": Hyslop correspondence with Hodgson, July, 7, 1900, ASPR.

257: his 1902 book: William James, *Varieties of Religious Experience: A Study of Human Nature* (New York: Longman, Green, 1902).

258: "Only on 426 days of my life": Myers, *Fragments of Inner Life.*

260: "up to the time": For the continued story of the Widow's Mite, see Funk, *Widow's Mite,* 155–77.

261: "A batch of reporters came after me": Hodgson to Hyslop, Apr. 7, 1903, beginning of an exchange of letters during that month, ASPR.

264: "ill-defined relations of the subliminal": Henry James, "Review of *Human Personality and Its Survival of Bodily Death,* by Frederic W. H. Myers (1903)," *Proceedings of the Society for Psychical Research* 18 (1903): 22–33.

11. A Force Not Generally Recognized

267: "Sometimes, I can hardly wait": Berger, *Lives and Letters in American Parapsychology,* 31.

269: "Well, Lord Rayleigh": quoted in "The King," in *Edwardian England, 1901–1914,* ed. Simon Nowell-Smith (London: Oxford University Press, 1964), 21; Barrett's presidential address cited in Haynes, *Society for Psychical Research,* 185.

269: "described in mocking detail": "Pepper Séance a Lively One," *New York Times,* Feb. 27, 1905; "General Montcalm's Crown Bobs up at a Séance," *New York Times,* Mar. 6, 1905.

271: "all the heroic qualities": WJ to Flournoy, Oct. 11, 1904, Houghton.

271: "I didn't at all": WJ to Hyslop, Nov. 11, 1904, Houghton.

271: "I have had a hard enough task": Hyslop to WJ, Feb. 27, 1905, ASPR.

272: "My Excellent Hyslop": WJ to Hyslop, Feb. 28, 1905, Houghton and ASPR.

273: "Go for the scoundrel": *Proceedings of the Society for Psychical Research* 14 (1899): 367.

273: "Absolutely sudden, dropt dead": WJ to Schiller, Jan. 16, 1909, Houghton.

273: Hodgson's funeral: Description of funeral from a report on "semi-annual" history of the Tavern Club, given by Dr. M. A. DeWolfe Howe, on May 7, 1906, and from a letter from WJ to Horace Fletcher on Dec. 25, 1905, Houghton, which talks about the number of people in tears.

274: "What is the matter?": This and all other such sittings that involve the reputed return of Richard Hodgson's spirit from William James, "Report on Mrs. Piper's Hodgson-Control," *Proceedings of the Society for Psychical Research* 23 (1909): 2–121.

274: WJ to Flournoy, Feb. 9, 1906, letter contains prediction on ASPR future, Houghton.

275: "You lack the discretion": WJ to Hyslop, Feb. 7, 1906, Houghton and ASPR.

276: the cross-correspondence study: Material throughout this chapter about the cross-correspondence studies is drawn from J. G. Piddington, "Concordant Automatisms", *Proceedings of the Society for Psychical Research* 22 (1908); Flournoy, *Spiritism and Psychology,* 174–85; Piper, *Life and Work,* 129–68; Tyrrell, *The Personality of Man,* 145–50; Conan Doyle, *History of Spiritualism,* 2:85–93; and Heywood, *The Sixth Sense,* 69–112.

281: "Pepper went on trial": "Mrs. Pepper a Bride; To Stop Spookfests," *New York*

Times, June 5, 1907; "Vanderbilt Signed Checks for Spook," *New York Times,* June 14, 1907; and "Mrs. Pepper Heard on Witness Stand," *New York Times,* Sept. 6, 1907.

286: records of the sittings: James, "Mrs. Piper's Hodgson-Control."

287: Everard Feilding: Feilding is profiled in the introduction to *Sittings with Eusapia Palladino and Other Stories,* a collection of Feilding's early-twentieth-century psychical research reports (Hyde Park, N.Y.: University Books, 1963); his description of the London medium can be found on pp. x–xi. The accounts of the Palladino sittings are taken from Feilding, Baggally and Carrington, "Sittings with Eusapia Palladino."

293: "The paramount importance": Barrett, *Threshold of the Unseen,* 1–9.

12. A Ghost Story

296: "his analysis of Mrs. Piper's Hodgson personality": James, "Mrs. Piper's Hodgson-Control."

300: "few people who looked into the evidence": Nora Sidgwick lecture, Jan. 25, 1912, which expanded on presidential address, reprinted in Sidgwick, *Mrs. Henry Sidgwick,* 301–11.

303: "perish in their ignorance and conceit": WJ to Carrington, June 15, 1909; in *Letters to Hereward Carrington* (privately printed, 1957), 41.

303: "Seriously to investigate": G. Stanley Hall, introduction to Tanner, *Studies in Spiritism.* All accounts of the Tanner and Hall studies of Mrs. Piper are taken from this book. For accounts of injuries and the response of Alta and Leonora Piper, see Piper, *Life and Work,* 173–75.

305: "By this morning's post": Lodge's exchange with Hall detailed in letters to WJ on Nov. 9, 1909, and Dec. 14, 1909, Houghton.

307: "Poor Carrington": TJ to Flournoy, Sept. 28, 1909, Houghton.

309: "Alice would barely speak to Munsterberg": WJ to William James Jr., Feb. 27, 1903, Houghton.

309: "Were it not for my fixed belief": WJ to Munsterberg, June 28, 1906, Houghton

309: "insinuate that I also am one": WJ to Flournoy, Jan. 26, 1910.

309: "Fraud with the feet": Flournoy to WJ, Mar. 15, 1910.

310: "The Confidences of a Psychical Researcher": William James, "The Confidences of a Psychical Researcher," *American Magazine* 68 (Oct. 1909): 580–89.

311: "Carrington had tried to repair her reputation": Carrington's account of the American sittings and his efforts to control damage afterward can be found in *Personal Experiences in Spiritualism* (London: J. Werner Laurie, Ltd., 1913).

312: "Everything this time was different": *Proceedings of the Society for Psychical Research* 25 (1911): 57–69.

312: "Man's character is too sophistically mixed": James, "Confidences of a Psychical Researcher."

314: "I came in with Halley's comet": "Mark Twain: A Look at the Life and Works of Samuel Clemens," www.hannibal.net/twain/biography/.

314: Podmore's death is described in Eric Dingwall's introduction to *Mediums of the 19th Century,* xxi–xxii.

315: "Suicide has . . . been suggested": Piddington to Hyslop, Nov. 1, 1910, ASPR.

316: "He wanted to go": Alice James to Pauline Goldmark, Sept. 14, 1910, Houghton.

316: "I believe in immortality": Simon, *Genuine Reality,* 385.

317: "Human Beings Only an Aggregate of Cells": Edward Marshall, "'No Immortality of the Soul' says Thomas A. Edison," *New York Times,* Oct. 2, 1910.

319: "Our duty is plain": Richet, *Thirty Years,* 625.

319: "The public is what it is": Feilding to Carrington, Aug. 15, 1912, in *Letters to Hereward Carrington,* 19.

320: "Any man who does not accept": Introduction to James H. Hyslop, *Contact with the Other World* (New York: Century Company, 1919).

320: "unwillingly children of the time": E. Feilding, "Can Psychical Research Contribute to Religious Apologetics?" *Dublin Review,* Apr.–June 1925; reprinted in Feilding, *Sittings with Eusapia Palladino,* 326–334.

INDEX

FOR THE BEST IN PAPERBACKS, LOOK FOR THE

In every corner of the world, on every subject under the sun, Penguin represents quality and variety—the very best in publishing today.

For complete information about books available from Penguin—including Penguin Classics and Puffins—and how to order them, write to us at the appropriate address below. Please note that for copyright reasons the selection of books varies from country to country.

In the United States: Please write to *Penguin Group (USA), P.O. Box 12289 Dept. B, Newark, New Jersey 07101-5289* or call 1-800-788-6262.

In the United Kingdom: Please write to *Dept. EP, Penguin Books Ltd, Bath Road, Harmondsworth, West Drayton, Middlesex UB7 0DA.*

In Canada: Please write to *Penguin Books Canada Ltd, 90 Eglinton Avenue East, Suite 700, Toronto, Ontario M4P 2Y3.*

In Australia: Please write to *Penguin Books Australia Ltd, P.O. Box 257, Ringwood, Victoria 3134.*

In New Zealand: Please write to *Penguin Books (NZ) Ltd, Private Bag 102902, North Shore Mail Centre, Auckland 10.*

In India: Please write to *Penguin Books India Pvt Ltd, 11 Panchsheel Shopping Centre, Panchsheel Park, New Delhi 110 017.*

In the Netherlands: Please write to *Penguin Books Netherlands bv, Postbus 3507, NL-1001 AH Amsterdam.*

In Germany: Please write to *Penguin Books Deutschland GmbH, Metzlerstrasse 26, 60594 Frankfurt am Main.*

In Spain: Please write to *Penguin Books S. A., Bravo Murillo 19, 1° B, 28015 Madrid.*

In Italy: Please write to *Penguin Italia s.r.l., Via Benedetto Croce 2, 20094 Corsico, Milano.*

In France: Please write to *Penguin France, Le Carré Wilson, 62 rue Benjamin Baillaud, 31500 Toulouse.*

In Japan: Please write to *Penguin Books Japan Ltd, Kaneko Building, 2-3-25 Koraku, Bunkyo-Ku, Tokyo 112.*

In South Africa: Please write to *Penguin Books South Africa (Pty) Ltd, Private Bag X14, Parkview, 2122 Johannesburg.*